Lecture Notes in Biomathematics

Managing Editor: S. Levin

35

Uwe an der Heiden

Analysis of Neural Networks

Springer-Verlag
Berlin Heidelberg New York 1980

Author

Uwe an der Heiden
Universität Bremen, FB 3
Postfach 330440
2800 Bremen 33
Federal Republic of Germany

AMS Subject Classifications (1980): 92A05

ISBN-13: 978-3-540-09966-6 e-ISBN-13: 978-3-642-45517-9
DOI: 10.1007/978-3-642-45517-9

Library of Congress Cataloging in Publication Data. Heiden, Uwe an der, 1942- Analysis of neural networks. (Lecture notes in biomathematics; 35) Bibliography: p. Includes index.
1. Neural circuitry--Mathematical models. I. Title. II. Series. QP363.3.H44 599.01'88
80-12023

2141/3140-543210

To DORIS

PREFACE

The purpose of this work is a unified and general treatment of activity in neural networks from a mathematical point of view. Possible applications of the theory presented are indicated throughout the text. However, they are not explored in detail for two reasons : first, the universal character of neural activity in nearly all animals requires some type of a general approach; secondly, the mathematical perspicuity would suffer if too many experimental details and empirical peculiarities were interspersed among the mathematical investigation. A guide to many applications is supplied by the references concerning a variety of specific issues.
Of course the theory does not aim at covering all individual problems. Moreover there are other approaches to neural network theory (see e.g. Poggio-Torre, 1978) based on the different levels at which the nervous system may be viewed.
The theory is a deterministic one reflecting the average behavior of neurons or neuron pools. In this respect the essay is written in the spirit of the work of Cowan, Feldman, and Wilson (see sect. 2.2).

The networks are described by systems of nonlinear integral equations. Therefore the paper can also be read as a course in nonlinear system theory. The interpretation of the elements as neurons is not a necessary one. However, for vividness the mathematical results are often expressed in neurophysiological terms, such as excitation, inhibition, membrane potentials, and impulse frequencies.
The nonlinearities are essential constituents of the theory. Important phenomena such as hysteresis, limit cycles, pulses only occur with nonlinear systems, and obviously neurons are nonlinear elements.

The term "analysis" in the title is meant in the mathematical sense. In particular the statements marked as theorems are provable. Of course many interesting and important properties of nonlinear systems can only be investigated by numerical methods. On the other hand there is now rapid progress in nonlinear analysis, and neural network theory should profit from this development.

Proofs of theorems are only given when the author could not find a result in the literature including the theorem.
The book is organized as follows. In the first chapter the network equations are derived on the basis of the universal properties and interaction principles of neurons and their organization in neural tissues. The second chapter shows how our model is related to other models in the literature, in particular to some experimentally investigated networks. The analysis is started in chap.3 establishing mathematical conditions in order that the equations unambiguously determine the states of the (model) networks. All further chapters investigate the (temporal or stationary) behavior of the model-networks following from the equations. Chapters 4-6 are restricted to nets composed of finitely many elements (these may be single neurons or populations of neurons with spatially homogeneous behavior). The last three chapters exhibit modes of behavior necessarily dependent on the spatial extension of neural tissues, such as wave form activity or spatial patterns. The last section (9.4) establishes some connexions to other literature and applications.
Essential parts of chapters 1-3 appeared in my paper "Structures of excitation and inhibition" published in Vol.21 of this series. Section 6.3 contains a slight generalization of a result which appeared in the Journal of Mathematical Analysis and Applications. Chapters 1-6 essentially coincide with my "Habilitationsschrift" submitted to the Faculty of Biology at the University of Tübingen in January 1979.

ACKNOWLEDGEMENT

I take the opportunity to thank Prof.Dr.K.P.Hadeler for his constant personal and scientific support during the years I have been one of his coworkers at the Lehrstuhl für Biomathematik of the University of Tübingen, where the book was written. I also thank my collegues, A.Wörz, K.Schumacher, F.Rothe, for many stimulating discussions. K.P.Hadeler and M.C.Mackey helped by many comments for improving the manuscript. Finally I am indebted to Mrs.B.Pölter for kindly providing the typescript.

December 1979 Uwe an der Heiden

TABLE OF CONTENTS

1. The general form of a neural network

1.1. Introduction

The generation, the transport, and the processing of information in the nervous system are achieved by the parallel activity of a large number of nerve cells. In the following a model as flexible as possible for the interaction of neurons and for the activity in neural networks will be developed. The model will be derived from the basic properties and interaction principles common to many types of nerve cells so far observed. A characteristic feature of the theory presented here is its formal generality. We do not provide a description of individual neurons or individual networks, but a framework, i.e. the general form of a neural network. Special networks are obtained by specifying the functions and parameters in the framework either by experimental measurements or by derivation from other more specific neuron models and interaction principles.

In particular it will turn out that many of the classical and more recent models are special cases of our theory. There are at least two advantages of this general approach. First, an exact relationship between the models in the literature is established allowing continuous transition from one model to the other in a mathematical sense. Second, many results previously derived for each of the singular models may now be obtained at once for all of them.
However, the most essential property of the theory is its suitability for the incorporation of experimental data. In this way congruence with experience, excluded in many models because of their rigidity, may hopefully be achieved in an area distinguished by variability and diversity. Of course it is not claimed that all neurophysiological phenomena can be covered by the system.

1.2. The transformation of impulse frequencies into generator potentials (intercellular transmission)

In this section the general form of the interaction of two neurons A and B is determined. Neurons contact each other via synapses. There can be several synapses between two cells, more than 10 synapses is not exceptional. The influence of a synapse depends on its internal effectiveness, on its location in the dendritic tree of the postsynaptic cell, on the transmission properties from the locus of the synapse towards another locus (e.g. of the axon hillock) where the synaptic information may have some importance. In view of this complexity it seems impractical in modeling neural networks to make explicit the local transmission properties of single synapses. What matters and what seems to be within the range of experimental observation and mathematical description is the <u>total influence</u> of cell A on cell B. (For processing in dendritic trees we remember, but do not refer explicitly to the work of W.Rall, see also Barrett and Crill, 1974 b, Rinzel,1975).

Assume that neuron A is at rest for all time with the exception of a single spike (=nerve impulse =action potential) generated at the axon hillock of A at time t=0. This means the impulse frequency (=number of spikes per second) $x_A(t)$ in neuron A obeys

$$x_A(t) = \delta(t) \ \sec^{-1} \quad , \quad -\infty < t < \infty ,$$

where $\delta(t)$ denotes Dirac's function.

Moreover assume for $t<0$ the membrane potential at the soma near to the axon hillock of cell B (this is called the generator potential v_B of cell B) is equal to the resting potential v_B .
If there is no influence from other cells on cell B (even an influence of spikes in cell B, i.e. any kind of self-inhibition or self-excitation, is excluded here; these effects will be considered later on) then the response of v_B to the impulse in A is described by a function

$$v_B(t)=v_{Bo} + h_{BA}(t) \quad , \quad -\infty<t< \infty$$

with

$$h_{BA}(t)=0 \quad \text{for } t<0 \quad ,$$

see fig.1.1.

The influence of A on B is excitatory if $h_{BA}(t)\geq 0$ for all t.
It is inhibitory if $h_{BA}(t)\leq 0$ for all t. However, impulse

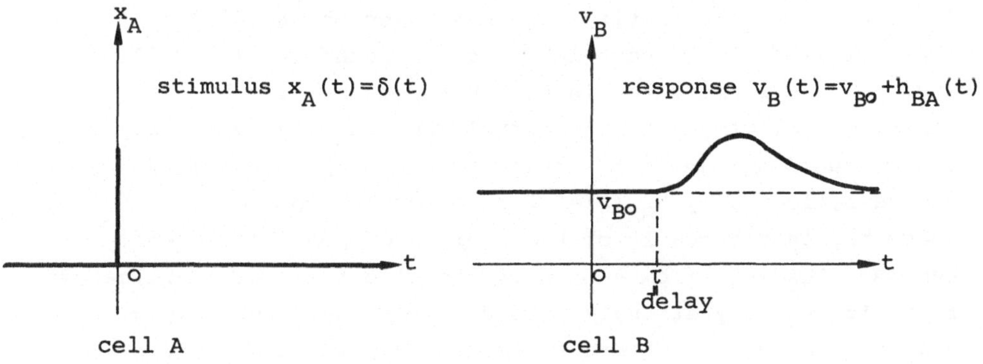

Fig. 1.1

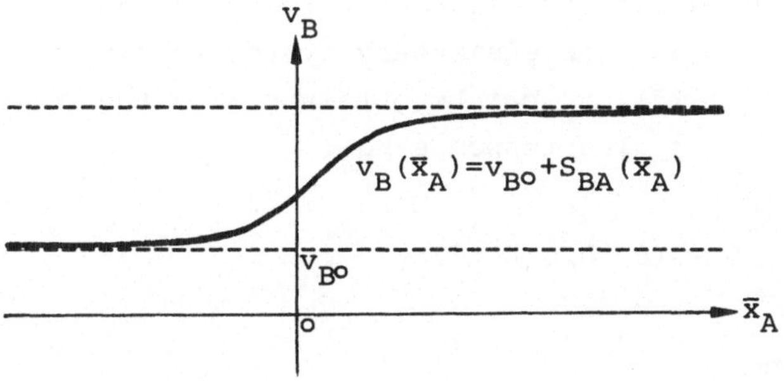

Fig. 1.2

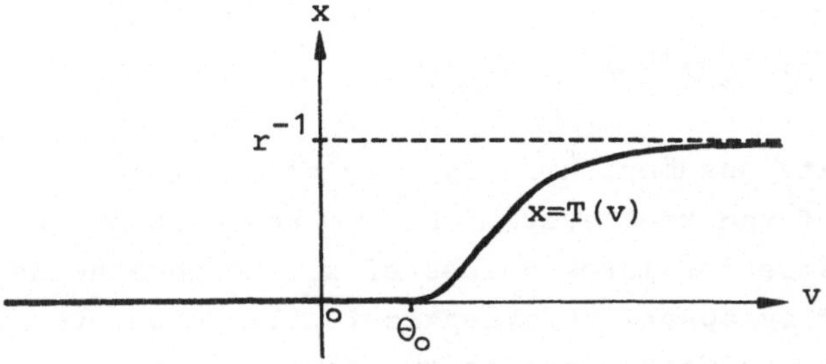

Fig. 1.4

responses with alternating sign have been observed also (Ratliff,1974). Thus post-inhibitory rebound (Perkel-Mulloney, 1974) can be represented in h_{BA} by a domain where h_{BA} is positive following a time interval where h_{BA} is negative. The conduction times for impulses along axons and synaptic delays can be reflected by $h_{BA}(t)=0$ for $0 \leq t < \tau$ with some $\tau > 0$. Since h_{BA} is the result of the overlap of possibly several synapses leading from A to B it can have a rather complicated form. It can vary strongly with different cell types (e.g. depending on whether there is active or passive dendritic transport).

For a moment it is supposed that the transformation of an arbitrary impulse frequency x_A of cell A into the generator potential v_B of cell B is that of a linear, time-invariant and causal system. Then, as an elementary system theory result (see e.g. Varjú, 1977), v_B can be obtained from the convolution of x_A with the impulse response h_{BA}:

$$v_B(t) = v_{Bo} + \int_{-\infty}^{t} x_A(t')h_{BA}(t-t')dt' \quad . \tag{1.1}$$

With the abbreviation

$$f*h(t) = \int_{-\infty}^{t} f(t')h(t-t')dt' \tag{1.2}$$

equ. (1.1) reads

$$v_B = v_{Bo} + x_A*h_{BA} \quad . \tag{1.3}$$

As a rule the linearity assumption leads only to a first order approximation of the real system. Due to the exhaustion of transmitter substance for large values of x_A, to thresholds for the activation of synapses, to synaptic facilitation, to postsynaptic saturation effects, and to the fact that the superposition of all synaptic influences is not strictly additive, the influence of x_A on v_B is nonlinear.

Obviously the nonlinear relation can be very complicated. In order to obtain reasonable network equations consider the following simplifying approximation.
Let $x_A=\bar{x}_A$ be constant (independent of time). Let $S_{BA}(\bar{x}_A)$ be the average generator potential produced in cell B by $\bar{x}_A$ alone. In order to measure the function S_{BA} effects of self-inhibition or self-excitation in cell B have to be excluded, i.e. spike generation in cell B has to be eliminated, at least theoretically. Then the steady state behavior is described by

$$v_B = v_{Bo} + H_{BA} \cdot S_{BA}(\bar{x}_A) \quad , \tag{1.4}$$

see fig. 1.2.
In general $S_{BA}: \mathbb{R}_+ \longrightarrow \mathbb{R}_+$ will be a nonlinear, monotone increasing, and bounded function. The factor H_{BA} is +1 or -1 according to whether the influence of A on B is inhibitory or excitatory respectively. The dynamic aspect in equ. (1.1) and the steady state aspect in equ. (1.4) are combined in the equation

$$v_B(t) = v_{Bo} + \int_{-\infty}^{t} S_{BA}(x_A(t'))h_{BA}(t-t')dt' \quad . \tag{1.5}$$

With the notation of (1.2) this equation is written

$$v_B = v_{Bo} + S_{BA}(x_A) * h_{BA} \quad . \tag{1.6}$$

Clearly the normalizing convention

$$H_{BA} = \int_0^\infty h_{BA}(t)dt$$

has to be observed. The form of equ. (1.5) suggests the following interpretation. $x_A(t)$ can be viewed as the presynaptic input of cell B received from cell A. The function $S_{BA}(x_A)$ represents a stationary non-linear weighting of the input by synaptic and dendritic mechanisms. The influence of $x_A(t')$, $t' \leq t$, on $v_B(t)$ is given by the product $S_{BA}(x_A(t'))h_{BA}(t-t')$. The influences for all $t' \leq t$ are summed up, which leads to equ. (1.5). Therefore the function h_{BA} appears as a temporal weighting factor. Its previous determination as an impulse response function represents an approximation which can be measured experimentally (at least in principle). It is proposed here that the nonlinear equation (1.5) is a reasonable approximation to the more general, but both mathematically and experimentally much more difficult model

$$v_B(t)=v_{Bo} + \int_{-\infty}^{t} K_{BA}(x_A(t'),t,t')dt'$$

with a nonlinear kernel K_{BA}.

A theoretical derivation of S_{BA} for a simple synaptic configuration is given in (Leibovic, 1972, chap.7), here S_{BA} is a sigmoid function. An experimental determination of S_{BA} for the squid giant synapse can be found in (Katz and Miledi, 1967, fig.9).

In many neurons the generation of a spike hyperpolarizes the generator potential of the same cell. This effect of the impulse frequency x_B of cell B on v_B is known as self-inhibition. For a moment it is assumed the effects of x_A and of x_B on v_B are separable at least theoretically and they obey the superposition principle. The effect of x_B on v_B can be viewed theoretically and experimentally just as the effect of x_A on v_B, leading to functions h_{BB} and S_{BB}. Including self-inhibition in this way, equ. (1.6) has to be modified to

$$v_B = v_{Bo} + S_{BA}(x_A) * h_{BA} + S_{BB}(x_B) * h_{BB} \; . \qquad (1.7)$$

The superposition principle will be weakened in sect.1.3.

Remark. It could happen that not all synaptic influences of cell A on cell B can be included in a single pair of functions (S_{BA}, h_{BA}). Then it may be possible to subdivide the synapses impinging from A on B into several groups, e.g. according to their location on different parts of the dendritic tree, each group having a description in form of equ. (1.3). This means that the cell A is replaced theoretically by cells $A_1, A_2, \ldots, A_n$, the number n corresponding to the number of groups, with pairs of functions (S_{BA_i}, h_{BA_i}), $i=1,2,\ldots,n$. All cells A_i have the same firing rate x_A. Equ. (1.7) generalizes to

$$v_B = v_{Bo} + S_{BB}(x_B) * h_{BB} + \sum_{i=1}^{n} S_{BA_i}(x_A) * h_{BA_i} \quad .$$

In the following it is assumed that this replacement has been performed already such that the individual cell is now A_i and not A.

1.3. The transformation of generator potentials into impulse frequencies (intracellular transmission)

In spike generating neurons at the beginning of the axon the generator potential is transformed into a series of impulses, which propagate along the axon.
A spike is generated if the generator potential v(t) exceeds a threshold value $\Theta(t)$ which itself depends on time. The variation of Θ with t can be derived theoretically from the models of Hodgkin and Huxley (1952) and Fitzhugh (1961, p. 449). However, the functions obtained from these models are only qualitatively correct. Quantitatively realistic values can only be determined by measurements. These measurements will vary greatly from cell type to cell type. The qualitative behavior of Θ is generally as follows:
Assume a spike is produced at the axon hillock at time $t_o=0$. Then during a certain interval $(0,r]$ of time, called the absolute

refractory period, the neuron is unable to generate a second spike, i.e. $\Theta(t)=\infty$ if $0<t\leq r$. Afterwards the threshold continuously decreases towards an asymptotic value Θ_0, the resting threshold, provided that no second impulse is initiated, i.e. that $v(t)<\Theta(t)$ for $t>r$, see fig.1.3.

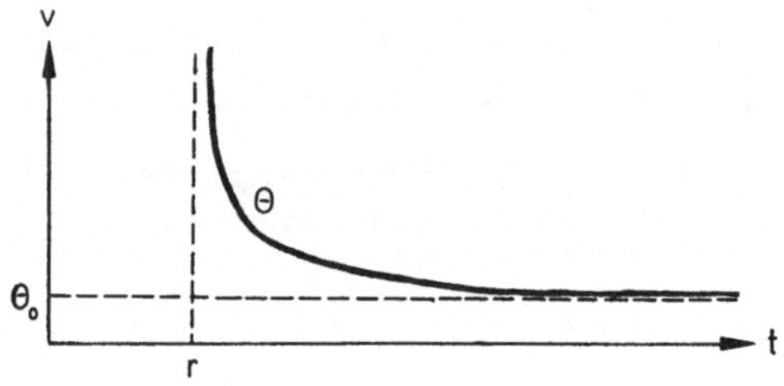

Fig.1.3. The threshold function Θ

The time during which Θ is finite and significantly larger than Θ_0 is called the relative refractory period. It varies between 1 and 200 msec with different cell types (Giesler and Goldberg, 1966). The first spike after t_0 occurs at the first time $t=t_1>t_0+r$ satisfying $v(t)\geq\Theta(t)$. It is assumed that after t_1 the threshold develops just as after t_0. A neuron is spontaneously active if its resting potential is smaller than its resting threshold, i.e. if $v_0<\Theta_0$.

All spikes of the same neuron have essentially the same shape. The details of the spikes do not contain any information. In general the information is contained in the impulse frequency (=number of spikes per second). Variations of the potential v within time intervals shorter than the sum of the duration of the spike and the absolute refractory period (about 2-20 msec) are not reflected by the impulse frequency. Therefore, in the following either it is assumed that the membrane potential does not vary too quickly within such intervals (the synonym slow potential indicates that this assumption is often satisfied) or the average of v with respect to such intervals has to be formed

("time course-graining", for details see Wilson-Cowan, 1972) leading to a new variable which will also be denoted by v. In both cases v will be called slow potential or generator potential.
Under these conditions the functional relationship between the potential v and the frequency x of a neuron follows from the temporal development of the threshold:

$$x=T(v)=\begin{cases} 0 & \text{if } v\leq\Theta_o \\ 1/\Theta^{-1}(v) & \text{if } v>\Theta_o \end{cases} \quad . \tag{1.8}$$

The symbol Θ^{-1} denotes the inverse function of Θ, i.e. $\Theta^{-1}(v)=t$ if $\Theta(t)=v>\Theta_o$.
Note that v has to be measured shortly before the beginning of the axon (the axon hillock) and x shortly behind the beginning of the axon.
$T: \mathbb{R} \longrightarrow \mathbb{R}_+$ is a nonnegative, monotone increasing, bounded function, its range being the interval $[0,r^{-1}]$, see fig.1.4. For real neurons it has to be determined experimentally since the threshold is not accessible to direct observation.
Let T_B denote the <u>potential to frequency conversion function</u> T for the cell B. Then it follows from the equs. (1.7), (1.8) that

$$x_B=T_B(v_{Bo} + S_{BA}(x_A) * h_{BA} + S_{BB}(x_B) * h_{BB}) \quad . \tag{1.9}$$

This equation has as variables only the impulse frequencies of the cells A and B (<u>impulse frequency representation</u>).
There is an equivalent <u>potential representation</u> using only the slow potentials v_A and v_B: Because of $x_A=T_A(v_A)$ and $x_B=T_B(v_B)$ it follows from (1.7) that

$$v_B=v_{Bo} + S_{BA}\circ T_A(v_A) * h_{BA} + S_{BB}\circ T_B(v_B) * h_{BB}, \tag{1.10}$$

(with the definition $f\circ g(\xi)=f(g(\xi))$ for two functions f,g).
Using the symbol

$$U_{BA} = S_{BA}\circ T_A \tag{1.11}$$

for the composed functions equ. (1.10) simplifies to

$$v_B = v_{Bo} + U_{BA}(v_A) * h_{BA} + U_{BB}(v_B) * h_{BB}, \tag{1.12}$$

which is the potential representation.
In general U_{BA} and U_{BB} are monotone increasing, nonnegative, and bounded functions. Whether the influences are excitatory or inhibitory depends on the sign of the temporal weight functions h_{BA} and h_{BB}.
In view of being a composed function U_{BA} can have a complex structure even if S_{BA} and T_A are simple sigmoid (=S-shaped) functions in the following sense.

Definition 1.1. Let $f: I \to \mathbb{R}$ be a continuous, monotone, and bounded function, defined on the interval $I=(a,b)$ and let n be a positive integer. f is called n-modal if there are numbers $a<\xi_1<\xi_2<\ldots<\xi_n<b$ such that

(i) on the intervals $I_1=(a,\xi_1)$, $I_2=(\xi_1,\xi_2),\ldots,I_n=(\xi_{n-1},\xi_n)$, $I_{n+1}=(\xi_n,b)$ the function f is convex or concave, and

(ii) on $I_i \cup I_{i+1}$ the function f is neither convex nor concave.

A 1-modal function is called sigmoid (typical graphs of sigmoid functions are shown in figs. 1.2,1.4).
Given any positive integer n it is possible to find two sigmoid functions f and g such that $g \circ f$ is n-modal. This fact, being important for the number of steady states in a neural network (see chap. 4), is illustrated by the following example.

Example 1.2 . Let the function f be defined on the interval $(0,\infty)$ by $f(\eta)=\sqrt{\eta}$. This function is nonnegative, monotone increasing, and concave. A second continuous nonnegative, monotone increasing, but convex function g is determined uniquely on $[0,\infty)$ by the conditions

$$g(0) = 0$$

$$g'(\xi) = \begin{cases} n & \text{for } \xi \in [2n-2,2n-1] \\ \xi-n+1 & \text{for } \xi \in [2n-1,2n] \end{cases} \quad , n=1,2,\ldots \ .$$

The function $h = g \circ f$ has infinitely many inflection points, i.e. it is ∞-modal. This will be proved by showing that the derivative of h has infinitely many local minima, separating convex and concave regions of h.
Let $a_n = (2n-2+1/2)^2$, $b_n = (2n-1+1/2)^2$ for $n = 1,2,3,\ldots$.
Note that $a_n < b_n < a_{n+1}$.

$$h'(\eta) = \frac{1}{2} g'(f(\eta))/\sqrt{\eta} \quad ,$$

$$h'(a_n) = \frac{1}{2} n/(2n-3/2), h'(b_n) = \frac{1}{2}(n+\frac{1}{2})/(2n-\frac{1}{2}) \; .$$

It follows

$$h'(a_n) > h'(b_n) < h'(a_{n+1}) \text{ for } n = 2,3,\ldots \; .$$

Since h' is continuous there is a local minimum of h' between a_n and a_{n+1}, Q.E.D.

Remark. In equ.(1.12) impulse frequencies do not appear. Therefore, under suitable conditions, this equation can be used as a model for the interaction of non-spiking neurons. Such neurons are wide-spread in the central nervous system.

1.4. Structures of neural networks

Consider a network composed of n neurons $A_1, A_2, \ldots, A_n$ all obeying the interaction- and transmission principles described in the previous sections. It is a plausible assumption that the potential v_i near to the axon hillock of cell A_i does not discriminate between influences of different cells A_k and A_j. This symmetry condition is satisfied by assuming that the influences of all cells add up and are then transformed by the potential-to-frequency conversion nonlinearity T. Then it follows from equ.(1.9) after replacing the indices A,B by indices i,j corresponding to the pair of neurons (A_i, A_j) that

$$x_i = T_i(v_{io} + \sum_{j=1}^{n} S_{ij}(x_j) * h_{ij} + \bar{S}_i(E_i) * \bar{h}_i), \; i = 1,2,\ldots,n. \qquad (1.13)$$

The term $\bar{S}_i(E_i) * \bar{h}_i$ stands for an external input to the neuron A_i. Thus if A_i is a sensory cell E_i is the sensory input and $\bar{S}_i$ corresponds to the Weber-Fechner law. The dynamical influence of the external input is given by the convolution of $\bar{S}_i(E_i)$ with the temporal weight function $\bar{h}_i = \bar{h}_i(t)$, which is obtained approximately as the response of v_i (after elimination of all other influences) to the impulse shaped input $E_i(t) = \delta(t)$.
By the relations

$$x_i = T_i(v_i), \qquad U_{ij} = S_{ij} \circ T_j, \tag{1.14}$$

the impulse frequency representation (1.13) is equivalent to the potential representation

$$v_i = v_{io} + \sum_{j=1}^{n} U_{ij}(v_j) * h_{ij} + \bar{S}_i(E_i) * \bar{h}_i, \quad i=1,2,\ldots,n. \tag{1.15}$$

In the systems (1.13) and (1.15) the individual neurons are viewed as discrete units. However, under some conditions they are considered more appropriately as members of a continuum. Such conditions may be:

(α) Mathematical reasons: The number n of neurons is extremely large, and the differences in the parameters of adjacent neurons (assuming some topological arrangement of the cells) are so small that it is justified mathematically to pass over from the discrete system to the corresponding continuous system,

(ß) physiological reasons: the population of neurons under consideration forms a homogeneous, densely packed and relatively isolated subsystem of the nervous system, such that the behavior of single neurons has no significance.
It will not be specified when (α) or (ß) is satisfied. Indeed, there are a considerable number of different cell types in the brain, the cells of the same type having very similar properties.

Moreover the cells of the same type are often arranged very regularly in the tissue, and the mutual interconnections between different types are also formed regularly. A splendid example is the various neural layers in the retina of the eye and in the visual cortex.
The advantage of a continuous representation compared with a discrete one consists in a more concise description and in a simplification of the mathematical analysis.

In line with these considerations assume there are given m types of neurons. The neurons of type k, k=1,2,...,m, are uniformly distributed in a region R_k of the nervous system. Geometrically R_k is interpreted as a 1-, 2- or 3-dimensional manifold according as the cells are arranged in a line, in a layer, or in a cluster. Each point $s_k \in R_k$ corresponds to a neuron, the impulse frequency and the membrane potential of which is denoted by $x(s_k,t)$ and $v(s_k,t)$ respectively. The potential-to-frequency transformation of the cell s_k is denoted by $T(s_k,v)$. Assuming the principles of the sections 1.2 and 1.3, just in the way as system (1.13) was obtained, the equations for the total neural network evolve as

$$x(s_k,t)=T[s_k,v_o(s_k)+\sum_{j=1}^{m}\int_{R_j} S(s_k,s_j',x(s_j',\cdot))*h(s_k,s_j',\cdot)ds_j'+ \tag{1.16}$$

$$+\tilde{S}(s_k,x(s_k,\cdot))*\tilde{h}(s_k,\cdot)+\bar{S}(s_k,E(s_k,\cdot))*\bar{h}(s_k,\cdot)] \quad ,$$

where $s_k \in R_k$, k=1,2,...,m.

The symbols $\tilde{}$ and $\bar{}$ refer to self-inhibition and to external inputs respectively. The dots indicate the places for the time variable. The convolution of $S(s_k,s_j',x(s_j',t))$ and $h(s_k,s_j',t)$ represents the influence of the cell $s_j' \in R_j$ on the cell $s_k \in R_k$. Introducing the functions

$$U(s_k,s_j',v)=S(s_k,s_j',T(s_j',v)),\tilde{U}(s_k,v)=\tilde{S}(s_k,T(s_k,v)), \tag{1.17}$$

and using the relation

$$x(s_k,t)=T(s_k,v(s_k,t)) \tag{1.18}$$

the equivalent potential representation of the network (1.16) is

$$v(s_k,t)=v_o(s_k)+\sum_{j=1}^{m}\int_{R_j} U(s_k,s_j',v(s_j',\cdot))*h(s_k,s_j',\cdot)ds_j'+$$
$$+\tilde{U}(s_k,v(s_k,\cdot))*\tilde{h}(s_k,\cdot)+\bar{S}(s_k,E(s_k,\cdot))*\bar{h}(s_k,\cdot) \tag{1.19}$$

for $s_k \in R_k$, $k=1,2,\ldots,m$.

Homogeneous interconnected layers of neurons falling within the scope of the equs.(1.16) have been investigated by many authors under various aspects (see e.g.Beurle,1956,Griffith,1963/65,von Seelen,1968/72,Marko 1969,Wilson-Cowan,1972/73,Cowan-Ermentrout 1978;see also end of sect.2.2).Some important cases will be indicated in the following chapter.

2. On the relations between several models for neural networks

This chapter will show that many of the classical and recent mathematical and experimental nerve net structures are contained in the general scheme developed in chap.1. In this way a well defined relation between all these different networks is established. A continuous transition from one model to the other is supplied implicitly. Results obtained for special networks can be applied or generalized to broader classes, and a proof for the general model suffices to deal with many special cases. Finally, the experimental and theoretical foundations of the specific networks often differ from that given for the general system in chap.1 ,thus enlarging its range of validity as a superior structure, independent from its foregoing derivation.

2.1. The retinal network of *Limulus polyphemus*; the Hartline-Ratliff equations

One of the experimentally and theoretically best investigated neural networks is contained in the retina of the complex eye of the arthropode *Limulus polyphemus* (horseshoe crab). The experiments predominantly performed by the school of Hartline and Ratliff show that the eccentric cells extending from the facets (ommatidia) to the optic nerve are interconnected by collaterals and that activity in one fiber decreases activity in neighboring fibers (lateral inhibition), (Ratliff, 1974, collected papers). If the cells are numbered from 1 to n, then in good agreement with the measurements the impulse frequencies $x_i(t)$, $i=1,2,\dots,n$, obey the system (1.13) with the following specializations (compare Ratliff-Knight-Graham, 1969)

$$
\begin{aligned}
T_i(v) &= b_i m(v-r_i) \quad ,\\
S_{ij}(x) &= K_{ij} m(x-r_{ij}),\\
h_{ij}(t) &= H(t-\tau)\,\delta_1 e^{-(t-\tau)/\delta_1} \qquad \text{for } i \neq j \\
h_{ii}(t) &= H(t)\,\delta_2 e^{-t/\delta_2} \quad ,
\end{aligned}
\tag{2.1}
$$

$i,j=1,2,\dots,n$. Here $b_i > 0$, $K_{ij} \leq 0$ (inhibitory coefficients), $r_i, r_{ij} \geq 0$ (threshold values), and δ_1, $\delta_2 > 0$ (time constants for lateral and self-inhibition respectively) are independent of time. The lateral inhibition acts with a delay of about $\tau = 0.1$ sec.

H denotes Heaviside's function

$$H(t) = \begin{cases} 0 & t < 0 \\ 1 & t \geq 0 \end{cases} .$$

The system (1.13) with (2.1) is nonlinear because of the function m defined by

$$m(x) = \begin{cases} 0 & x \leq 0 \\ x & x \geq 0 \end{cases} .$$

In stationary conditions ($x_i(t)=\bar{x}_i$=const., $E_i(t)=\bar{E}_i$=const.) the system (1.13) with (2.1) reduces to the well-known Hartline-Ratliff equations

$$\bar{x}_i=b_i\ m(e_i - \sum_{j=1}^{n} K_{ij}\ m(\bar{x}_j-r_{ij})),\quad i=1,2,\dots,n, \tag{2.2}$$

where $e_i=v_{io}+\bar{E}_i$ essentially measures the excitation of the membrane potential by the incoming light.
The systems (1.13) & (2.1) and (2.2) have been investigated theoretically and mathematically in many respects (see e.g. Reichardt-MacGinitie (1961,1962), Varju (1962,1965), Coleman-Renninger (1971, 1974, 1975, 1976), Hadeler (1974, 1977)). Part of the results will be referred to in the following chapters.

2.2. A statistical approach: Activities in coupled neuron pools; models of Cowan, Feldman, Wilson

According to their derivation the integral equations of chap. 1 describe the activities of single neurons within the cooperative structure of a neural network. However, histological findings suggest that in some types of nervous tissue (e.g. in intralaminar thalamus, in the brain stem, in the reticular formation, and in motoneuron pools in the spinal cord) it is not the individual cell which is functionally important but the activity pattern of large cell aggregates being itself composed of a number of local, relatively homogeneous pools of densely interconnected neurons. The only meaningful experimental approach to such pools is to measure the averaged evoked potentials and to determine post-stimulus-time histograms for the mean impulse activity.

From a theoretical point of view some kind of statistical consideration is necessary for the understanding and appropriate description of the global activity in neuron pools. Such an approach was given in a series of papers by Cowan, Wilson, and Feldman (1972-76).

It turns out that their models are formally a special case of the equations in chap. 1, though the interpretation of the variables and parameters differs.

A neuron pool is a local neural net consisting of possiby several types of neurons with randomly distributed interconnections depending on the distance of the cells.

For a detailed derivation of the following equations see (Feldman and Cowan, 1975). Given n local aggregates r_i of neurons. Let $A(r_i, r_j) = A_{ij}$ be the density of connections from neurons in r_j to neurons in r_i. The local spatial average $F_i = F(r_i, t)$ of the firing density in r_i is experimentally related to the post-stimulus-time histogram. As an average F_i is a smooth function of the average membrane potential V_i, which is experimentally measured by the averaged evoked potential:

$$F_i = S_i(V_i) \quad .$$

S can be approximated by

$$S(V) = \left[r - \tau \ln \frac{V - \vartheta_{TH}}{\vartheta_o - \vartheta_{TH}} \right]^{-1}$$

where r denotes the relative refractory period (=RFP), τ the time constant of the RFP, ϑ_{TH} the resting threshold, ϑ_o the threshold at the beginning of the RFP.

If the postsynaptic potential has the form $e=e_o \exp(-t/\mu)$ with a membrane time constant μ (of the order of a few tens of milliseconds) then the large scale activity in the neural net is determined by the system

$$\mu \frac{dF_i}{dt} = - F_i + S_i(\mu e_o \sum_j A_{ij} F_j + P_i), i=1,2,\ldots,n, \qquad (2.3)$$

P_i representing the external input to r_i.
Formal integration shows that (2.3) is formally a special case of system (1.13) (with $h_{ij}=h = e_o \exp(-t/\mu)$).Note that self-inhibitory or self-excitatory terms occur in (2.3) if there is lateral inhibition or excitation within the same aggregate respectively.Cowan and Feldman also derive field equations of the form

$$\mu \frac{\partial F(r,t)}{\partial t} = - F + \mathcal{S}[\mu . e_o \int_R A(r,r',t)F(r',t)+P(r,t)] \quad , \qquad (2.4)$$

which are special cases of the system (1.16) with the functions h all being of exponential type with a common time constant μ.
The model (2.3) is applied to the population of α-motoneurons innervating the external intercostal muscles of the rib cage (Feldman and Cowan 1975 a). Here P corresponds to the weighted sum of influences from intrafusal fibers projecting from sensory muscle spindles and supraspinal respiration centres in the pons and in the medulla. In (Feldman and Cowan 1975 b) a model of the form (2.3) accounts for sustained rhythmic activities of cell populations in the brain stem controlling the respiratory system.

Field equations of the form (2.4) and closely related types have been derived and investigated by Beurle (1956), Griffith (1963), v.Seelen(1968), Wilson and Cowan (1972,1973). Matrix equations similar to (2.3) and all contained in system (1.13) have been considered by Cowan (1967,1970), Aplevich (1967), Morishita and Yajima (1972), Stein et.al.(1974), Oguztöreli (1975), Tokura and Morishita (1977), Wevelsiep (1977).

It should be mentioned that the statistical approach of Cowan, Feldman and Wilson for the large scale activity of neuron populations can be generalized in order to give a broader class of models, comparable to that in chap. 1 of our discussion, and to extend the range of possible applications.

2.3. Discrete models

a) The logical neurons of McCulloch and Pitts

In contrast to the previous models in the classical model of McCulloch and Pitts (1943) the time is discretized into entire multiples $t=0, \tau, 2\tau, 3\tau, \ldots$ of a unit time τ which is in the order of magnitude of 1 msec.
The "logical" neurons are either in the state "0" (resting) or in the state "1" (firing). In a finite network the state $x_i(t)$ of neuron i, $i=1,2,\ldots,n$, at time t is determined by the state of the network at time $t-\tau$ and the external input $E_i(t-\tau)$:

$$x_i(t)=H(E_i(t-\tau)+\sum_{j=1}^{n} K_{ij}x_j(t-\tau)-\Theta_i), \qquad (2.5)$$

where the K_{ij}'s are interaction constants and Θ_i constant threshold values, H=Heaviside's function.
The equations (2.5) are related to the system (1.13) by specializing

$$T_i(v)=H(v-\Theta_i), S_{ij}(x_j)=K_{ij}x_j,\ h_{ij}(t)=\delta(t-\tau)=\bar{h}_i(t), \bar{S}_i(E_i)=E_i. \qquad (2.6)$$

Note that the empirical meaning of the variables is again different from the derivation in chap. 1.
Within the framework of system (1.13) there is a continuous transition from the McCulloch and Pitts model to the Hartline-Ratliff equations (characterized by (2.1)).

In order to make this explicit we assume (only for simplicity) that in (2.1) we have $K_{ii}=0$ and $r_{ij}=0$. For every $\lambda > 0$ a specialization of system (1.13) is defined by

$$T_i^{(\lambda)}(v)=\begin{cases} \lambda b_i \, m(v-\Theta_i) & \text{if } v<\Theta_i+1/(\lambda b_i) \\ 1+b_i(v-\Theta_i-1/(\lambda b_i))/\lambda & \text{if } v\geq\Theta_i+1/(\lambda b_i), \end{cases} \tag{2.7}$$

$$v_{io}=0,\ S_{ij}(x_j)=K_{ij}x_j,\ h_{ij}^{(\lambda)}(t)=H(t-\tau)\,\delta_1\, e^{-(t-\tau)/(\delta_1\lambda)} \quad .$$

For $\lambda=1$ the system (2.1) of Hartline-Ratliff is obtained. If $\lambda \longrightarrow \infty$ the system (1.13) with (2.7) converges to the model (2.5) of McCulloch and Pitts (note that in this formal transition the dimensions of the variables are not preserved). The discrete model has been extensively studied, see e.g. Caianiello et.al. (1961, 1967, 1976), Arbib (1965).
Remark. Each finite automaton can be described in the form (2.5) (see Arbib, 1965).

b) Discrete time and continuous states

In view of the fact that in many neurons the postsynaptic potential decays with a time constant μ in the order of magnitude of a few tens of a millisecond (Oshima, 1969) and that often a comparatively long time τ_{ij} elapses between the generation of a signal in a cell j and the response to the signal in the target cell i it may have sense to replace the "true" impulse response function

$$\tilde{h}_{ij}(t)=\frac{1}{\mu}H(t-\tau_{ij})e^{-(t-\tau_{ij})/\mu}$$

by the approximating function

$$h_{ij}(t) = \delta(t-\tau_{ij}) \quad .$$

If, moreover, the delays τ_{ij} are all nearly equal, $\tau_{ij} \approx \tau$, then according to the equations (1.13) the state of the network at time $t+\tau$ depends only on the state at time t and on the input $E_i * \bar{h}_i(t+\tau)$. In this situation the integral equations (1.13) reduce to a system of difference equations

$$x_i^{(\nu+1)} = T_i(v_{io} + \sum_{j=1}^{n} S_{ij}(x_j^{(\nu)}) + e_i^{(\nu)}), \qquad (2.8)$$

$i=1,2,\ldots,n;\ \nu=0,1,2,\ldots$
with the identifications

$$x_i^{(\nu)} = x_i(t_o + \nu\cdot\tau),\quad e_i^{(\nu)} = E_i * \bar{h}_i(t_o+\nu\cdot\tau) \quad .$$

In contrast to the logical neurons of McCulloch and Pitts, the states of these neurons have a continuous range.
As an example remember that, as mentioned in sect. 1 of this chapter, there is a delay $\tau \approx 0.1$ sec in the interaction of two ommatidia in the eye of <u>Limulus</u>. Indeed, Ratliff, Hartline and Miller (1963) proposed the system

$$x_i^{(\nu)} = m(e_i^{(\nu)} - \sum_{j=1}^{n} K_{ij}\, m(x_j^{(\nu)} - r_{ij})),\ i=1,2,\ldots,n. \qquad (2.9)$$

as qualitatively compatible with their experiments on lateral inhibition in the Limulus retina.
Hadeler (1974) investigated stability properties of system (2.9). Coleman and Renninger (1975) studied (2.9) under spatial homogenity conditions ($e_i^{(\nu)}=e$, $x_i^{(\nu)} = x^{(\nu)}$, $r_{ij}=r$), however admitting that $K_{ij} = K(e)$ depends on the excitation e.

3. Existence and uniqueness of time dependent solutions

Preceeding to the investigation of the properties of neural networks described by the integral equation systems (1.13) or (1.16) the question has to be settled whether by these equations for each point of time a unique state of the network is determined. Difficulties can be encountered if the model contains a contradiction. In this case the system (1.13) (or (1.16)) has no solution (existence problem). Another difficulty arises if the equations do not give a complete description of the network. Then the equations allow more than one solution and it is left open which of the solutions corresponds to the state of the network (uniqueness problem). Finally it could happen that the system determines a unique solution, but this solution cannot be extended beyond a certain finite time t^*. Then it remains unclear how the network will evolve after t^* (extension problem).

The following theorems give conditions sufficient for the existence and uniqueness of solutions to the systems (1.13) and (1.16) (and by that also for the equivalent systems (1.15) and (1.19) resp.).

Considering first the system (1.16), let $R = \dot{\bigcup}_{1 \leq k \leq m} R_k$ be the topological sum of the manifolds $R_1, R_2, \ldots, R_m$. An <u>initial condition</u> of system (1.16) is a continuous and bounded function $x: R \times (-\infty, 0) \rightarrow \mathbb{R}$. A <u>solution</u> of system (1.16) corresponding to the initial condition $x: R \times (-\infty, 0) \rightarrow \mathbb{R}$ is a function $x: R \times [0, \infty) \rightarrow \mathbb{R}$ satisfying the equations (1.16) for all $t \geq 0$ if the initial condition is inserted in these equations.

<u>Theorem 3.1</u>. Corresponding to any initial condition the system (1.16) of integral equations has a unique continuous and nonnegative solution if the following six conditions are satisfied for all $k, j = 1, 2, \ldots, m$:

(i) R_k is a bounded, closed $d(k)$-dimensional submanifold of a finite dimensional Euclidean space, R_k has a finite $d(k)$-dimensional volume,

(ii) $E(s_k, t) = 0$ for $t < 0$, $E(s_k, t)$ is continuous on $R_k \times [0, \infty)$,

(iii) $v_o(s_k)$ is continuous on R_k,

(iv) the functions $T(s_k,w)$, $S(s_k,s_j',w)$, $\tilde{S}(s_k,w)$, and $\bar{S}(s_k,w)$ are continuous on $R_k \times \mathbb{R}$, $R_k \times R_j \times \mathbb{R}_+$, $R_k \times \mathbb{R}_+$, and $R_k \times \mathbb{R}$ respectively and they satisfy a global Lipschitz condition with respect to the variable w,

(v) $T(s_k,w)$ is nonnegative,

(vi) $h(s_k,s_j',t)$, $\tilde{h}(s_k,t)$, and $\bar{h}(s_k,t)$ are bounded on $R_k \times R_j \times \mathbb{R}_+$ and $R_k \times \mathbb{R}_+$ respectively, for each fixed t they are continuous with respect to (s_k,s_j'), and they are piecewise continuous with respect to t; moreover the integrals

$$\int_0^\infty |h(s_k,s_j',t)|\,dt, \quad \int_0^\infty |\tilde{h}(s_k,t)|\,dt, \quad \int_0^\infty |\bar{h}(s_k,t)|\,dt$$

are finite for each $s_k \in R_k$, $s_j' \in R_j$.

<u>Remark 3.2.</u> A function $f(u,w)$ satisfies a <u>global Lipschitz condition</u> with respect to w if and only if there is a constant $L > 0$ such that

$$|f(u,w)-f(u,\tilde{w})| < L\,|w-\tilde{w}|$$

for all (u,w), $(u,\tilde{w})$ in the domain of f.

<u>Proof of theorem 3.1 .</u> Let $|R_k|$ be the volume of R_k, let $K > 0$ be a bound for the functions $h,\tilde{h}$ and $\bar{h}$, moreover let ε be a number with

$$0 < \varepsilon < \left(L^2K\left(1+\sum_{k=1}^{m}|R_k|\right)\right)^{-1} . \tag{3.1}$$

Let $t_o \geq 0$. Assume there are uniquely determined functions $x(s_k,t)$, defined on $R_k \times (-\infty,t_o]$, continuous on $R_k \times [0,t_o]$, coinciding on $R_k \times (-\infty,0)$ with the initial condition, and satisfying the system (1.16) for all $t \in [0,t_o]$. This assumption holds for $t_o=0$ as a consequence of the conditions of the theorem. The theorem will be proved if $x(s_k,\cdot)$ can be extended uniquely to the interval $I=[t_o,t_o+\varepsilon]$ such that $x(s_k,t)$ is continuous and nonnegative on $R_k \times I$ and satisfies (1.16).

In order to prove the existence of such an extension an operator $\mathcal{T}: \mathcal{B} \to \mathcal{B}$ is defined on the Banach space

$$\mathcal{B} = \left\{ f: R \times I \to \mathbb{R} : \quad f \text{ is continuous} \right\}$$

(with norm $\|f\| = \max \left\{ |f(s,t)| : (s,t) \in R \times I \right\}$) in the following way:

Let $f \in \mathcal{B}$. Define $y(s_k,t)=x(s_k,t)$ for $t < t_o$ and $y(s_k,t)=f(s_k,t)$ for $t \in I$, $s_k \in R_k$. For $(s_k,t) \in R_k \times I$ let $(\mathcal{T} f)(s_k,t)$ be defined by the value which is obtained from the right hand side of equ. (1.16) by inserting $y(s_k,\cdot)$ and $y(s_j',\cdot)$ in place of $x(s_k,\cdot)$ and $x(s_j',\cdot)$ respectively. The continuity and boundedness conditions of the theorem and the initial condition imply that $\mathcal{T}f$ is defined and that $\mathcal{T}f \in \mathcal{B}$.

For any two functions $f,g \in \mathcal{B}$ the following estimates hold:

$$|(\mathcal{T}f)(s_k,t)-(\mathcal{T}g)(s_k,t)| \leq$$

$$\leq L\Big(\sum_{j=1}^{m} \int_{R_j} \int_{t_o}^{t} L \; |f(s_j',t')-g(s_j',t')| \cdot |h(s_k,s_j',t-t')| \, dt' ds_j' +$$

$$+ \int_{t_o}^{t} L \; |f(s_k,t')-g(s_k,t')| \cdot |\tilde{h}(s_k,t-t')| \, dt' \Big) \leq$$

$$\leq L^2K(t-t_o)\left(\sum_{j=1}^{m} |R_j| +1\right) \|f-g\| \leq \varkappa \ \|f-g\|$$

where $\varkappa = L^2K\varepsilon\left(\sum_{j=1}^{m} |R_j| +1\right)$.

The final estimate is independent of (s_k,t), therefore

$$\|\mathcal{T}f - \mathcal{T}g\| \leq \varkappa \ \|f-g\|$$

with a constant $\varkappa < 1$, i.e. the operator $\mathcal{T}$ is a contraction on $\mathcal{B}$. According to the Banach fixed point theorem $\mathcal{T}$ has a unique fixed point f_o, $\mathcal{T}f_o = f_o$. It follows from the definition of $\mathcal{T}$ that $x|_{R\times I} = f_o$ is a continuous extension of the solution $x|_{R\times(-\infty,t_o]}$. On the other hand every extension of x to the interval I which satisfies (1.16) is a fixed point of $\mathcal{T}$. Therefore the solution is unique. The solution x is nonnegative on $R\times[0,\infty)$ since $T(s_k,w)$ is nonnegative. Q.E.D.

A corresponding theorem for the system (1.13) can be proved by analogy. However, for such a finite net of n neurons we can introduce artificially m manifolds R_j, $j=1,2,\dots,m=n$, with R_j consisting of exactly one point. Then the following theorem is a special case of theorem 3.1.

Theorem 3.3. For $i,j=1,2,\dots,n$ let the continuous functions $T_i: \mathbb{R} \longrightarrow \mathbb{R}_+$, $S_{ij}: \mathbb{R}_+ \longrightarrow \mathbb{R}$, and $\bar{S}_i: \mathbb{R} \longrightarrow \mathbb{R}$ satisfy a global Lipschitz condition.
Let the functions $h_{ij}: \mathbb{R}_+ \longrightarrow \mathbb{R}$ and $h_i: \mathbb{R}_+ \longrightarrow \mathbb{R}$ be piecewise continuous, bounded, and integrable. Finally let $E_i: \mathbb{R} \longrightarrow \mathbb{R}$ be continuous on $\mathbb{R}_+$ and $E_i(t)=0$ for $t < 0$. Then for each continuous

and bounded initial condition the system (1.13) has a unique continuous solution. This solution is nonnegative if the continuous functions T_i are nonnegative.–

The conditions in the previous theorems are not the most general ones for the solvability of the integral systems. Thus the boundedness of the manifolds R_k can be dropped if the initial conditions are required to fall off sufficiently fast for large values of the space variables. Initial conditions and solutions are then elements of $\mathcal{L}^p$-spaces or other suitable Banach spaces. Dispensing with the continuity of the solutions the continuity conditions in theorem 3.1 can be considerably weakened. Eventually the global Lipschitz conditions can be replaced by local continuity and boundedness conditions for the functions S and T. All this will not be pursued here in order to avoid lengthiness.

4. Steady states of finite-dimensional networks

All equations so far considered describe dynamical systems and as such they have time dependent solutions. A solution which is constant with respect to time is called a steady state of the system. The main questions investigated in this chapter, which is restricted to systems of the form (1.13) or (1.15), are

a) do steady states exist?
b) how many steady states are there?
c) in which way do the steady states depend on the (constant) input?

These questions are very related to hysteresis problems in neural networks (see sect.4.2).

4.1. Existence problem

Steady states generally only exist if either there is no input to the network or the input is temporally constant.

Therefore it is assumed throughout this chapter that in the systems (1.13) and (1.15)

$$E_i(t) = \bar{E}_i = \text{constant} \quad .$$

The constant vector $(x_1(t), x_2(t), \ldots, x_n(t)) = (\bar{x}_1, \bar{x}_2, \ldots, \bar{x}_n)$ is a steady state solution of the system (1.13) if and only if

$$\bar{x}_i = T_i(e_i + \sum_{j=1}^{n} H_{ij} S_{ij}(\bar{x}_j), \quad i=1,2,\ldots,n \tag{4.1}$$

with the abbreviations

$$e_i = v_{io} + \bar{S}_i(\bar{E}_i) \cdot \int_0^\infty h_i(t)dt \quad , \tag{4.2}$$

$$H_{ij} = \int_0^\infty h_{ij}(t)dt \quad .$$

The numbers H_{ij} are normalized (by multiplying S_{ij} with an appropriate factor) in such a way that

$$H_{ij} \in \{-1, 0, 1\}$$

according as the influence of the neuron j on the neuron i is inhibitory, zero, or excitatory respectively. Consequently the functions $S_{ij} : \mathbb{R}_+ \longrightarrow \mathbb{R}_+$ are assumed to be nonnegative. Similarly $(v_1(t), v_2(t), \ldots, v_n(t)) = (\bar{v}_1, \bar{v}_2, \ldots, \bar{v}_n)$ is a steady state solution of the system (1.15) if and only if

$$\bar{v}_i = e_i + \sum_{j=1}^{n} H_{ij} U_{ij}(\bar{v}_j), \qquad i=1,2,\ldots,n. \tag{4.3}$$

The connection between the equivalent systems (4.1) and (4.3) is given by the transformations

$$U_{ij} = S_{ij} \circ T_j, \quad \bar{x}_i = T_i(\bar{v}_i), \qquad \bar{v}_i = e_i + \sum_{j=1}^{n} H_{ij} S_{ij}(\bar{x}_j) \quad . \tag{4.4}$$

The following theorem gives a condition sufficient for the existence of a steady state in the networks (1.13) and (1.15).

Theorem 4.1 . For arbitrary constants e_i and $H_{ij} \in \{-1,0,1\}$ the nonlinear system (4.1) has at least one solution $(\bar{x}_1,\ldots,\bar{x}_n)$ with $\bar{x}_i \geq 0$, $i=1,2,\ldots,n$, if the functions $T_i: \mathbb{R} \to \mathbb{R}_+$ and $S_{ij}: \mathbb{R}_+ \to \mathbb{R}_+$ are continuous and such that $U_{ij}=S_{ij} \circ T_j$ is bounded for $i,j=1,\ldots,n$.

Proof. The proof is based on Brouwer's fixed point theorem : Let $\Omega \subset \mathbb{R}^n$ be a bounded, closed, and convex set, and let $F:\Omega \to \Omega$ be a continuous function, then there is a fixed point $\omega \in \Omega$, $F\omega = \omega$. In order to apply this theorem define

$$M_i = \sum_{j=1}^{n} \sup_{v \in \mathbb{R}} U_{ij}(v) \quad ,$$

$$\Omega = \{(v_1,\ldots,v_n) \in \mathbb{R}^n : e_i - M_i \leq v_i \leq e_i + M_i, i=1,2,\ldots,n\}.$$

Then the map $F=(F_1,F_2,\ldots,F_n):\Omega \to \mathbb{R}^n$ defined by

$$F_i(v_1,\ldots,v_n) = e_i + \sum_{j=1}^{n} H_{ij} U_{ij}(v_j)$$

is continuous and satisfies

$$|F_i(v_1,\ldots,v_n)| \leq M_i \quad ,$$

which implies $F(\Omega) \subset \Omega$. Brouwer's fixed point theorem supplies an element $\bar{v}=(\bar{v}_1,\ldots,\bar{v}_n) \in \Omega$ with $F(\bar{v})=\bar{v}$. According to the definition of F the vector $\bar{v}$ is a solution of the system (4.3), and consequently the nonnegative vector $(\bar{x}_1,\ldots,\bar{x}_n)=(T_1 \bar{v}_1,\ldots,T_n \bar{v}_n)$ solves the system (4.1). Q.E.D.

4.2. The number of steady states

The following discussion will show that in general there are several steady states possible in a given neural network. This fact is the basis for the occurence of hysteresis phenomena in such nets. Since it is difficult to give general criteria for the number of solutions of nonlinear equation systems at first the special cases n=1 and n=2 will be investigated in some detail.

4.2.a) Single neurons

The equation for the stationary potential $\bar{v}$ of a single, isolated cell is (n=1 in equ. (4.3))

$$\bar{v}=e + H\ U(\bar{v}), \tag{4.5}$$

where $e=v_0+e_e$, v_0=resting potential, e_e=constant external input, and $U: \mathbb{R} \longrightarrow \mathbb{R}_+$ is (in general) a monotone increasing, bounded function.

For H=O (no self-inhibition or self-excitation) $\bar{v}=v_0+e_e, \bar{x}=T(\bar{v})$ are uniquely determined. If $T(v_0)> 0$ and $e_e=0$, in the steady state the cell is firing spontaneously with the frequency $T(v_0)$.

With self-inhibition (H=-1) the right hand side of equ. (4.5) is a monotone decreasing function of $\bar{v}$, and the left hand side is strictly increasing from - to .Therefore for every fixed value e there is exactly one potential $\bar{v}=\bar{v}(e\ )$ satisfying equ. (4.5). For increasing functions U the relation :

$$\text{if} \quad e < \tilde{e} \quad , \text{ then } \bar{v}(e\ ) < \bar{v}(\tilde{e}\ ) ,$$

holds.

Let now H=1.Self-excitation seems to be seldom realized in single neurons.However,since the equs. (4.3) also define steady states

of neuron pools (see sect.2.2) and of homogeneous populations of neurons (Coleman-Renninger 1974-1976)it has to be observed that self-excitation in this context means lateral excitation between neurons of the same species.
Equ.(4.5) can be written

$$\bar{v}-e = U(\bar{v}), \tag{4.6}$$

where $U: \mathbb{R} \rightarrow \mathbb{R}_+$ is assumed to be a monotone increasing, bounded function, which can be n-modal (see definition 1.1 and example 1.2). Theorem 4.1 implies that for each fixed input e there is at least one solution $\bar{v}$ of equ. (4.6). In the simplest case U(v) is sigmoid. Moreover it is assumed that U is a differentiable function and has a single inflection point at $\bar{v}=v_1$. If $U'(v_1)< 1$ then for each e the line $w=\bar{v}-e$ (= left hand side of equ. (4.6)) has exactly one intersection point with the curve $w=U(\bar{v})$ (=right hand side of equ. (4.6)). This means that to each input e there corresponds exactly one steady state $\bar{v}=\bar{v}(e)$. The steady state is a strictly increasing sigmoid function of the input.
However, if $U'(v_1) > 1$, then depending on the value of e there are one, two, or three solutions of equ. (4.6) as indicated in fig.4.1 (with $e=e_1$, $e=e_2$, $e=e_3$ resp.)

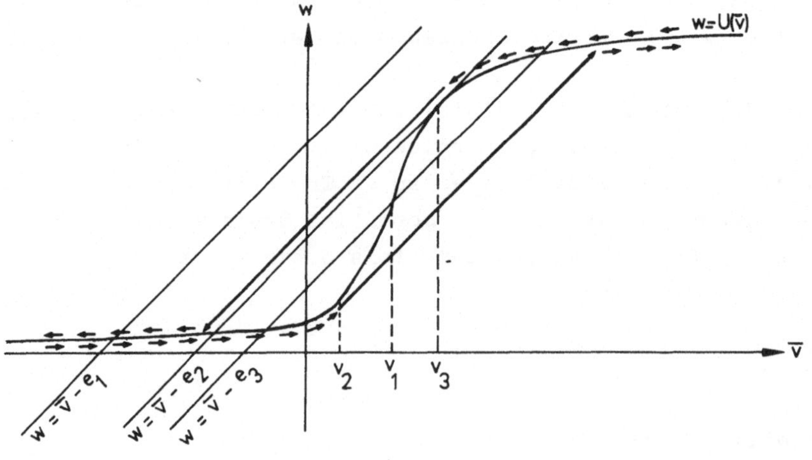

Fig.4.1

Let v_2 and v_3 denote the two values of $\bar{v}$, where $U'(\bar{v})=1$, $v_2 < v_3$. The two exceptional cases where there are exactly two steady states belong to $e=v_2-U(v_2)=\underline{e}$ and $e=v_3-U(v_3)=\bar{e}$.
For $\underline{e} < e < \bar{e}$ there are three steady states, for $e < \underline{e}$ or $e > \bar{e}$ there is only one steady state.
If e is sufficiently slowly increased (such that the system is always at equilibrium) from a value below $\underline{e}$ to a value above $\bar{e}$ and afterwards slowly decreased below $\underline{e}$ then the system runs through a hysteresis loop (as indicated by the arrows in fig.4.1).
More complex hysteresis phenomena can arise if the function U is n-modal with $n > 1$. Assume in the following that U is twice differentiable and that the second derivative U" only vanishes at the n inflection points $v_1 < v_2 < \ldots < v_n$ of U (this assumption is in a certain sense generic). Since U(v) is convex for $v < v_1$ and concave for $v > v_n$, the number n is odd.
If $U'(v) < 1$ for all v then equ.(4.6) has exactly one solution $\bar{v}=\bar{v}(e)$ for each e, namely

$$\bar{v}=(I-U)^{-1}(e),$$

where I denotes the identity function: $I(v)=v$. The inverse $(I-U)^{-1}$ of the function I-U exists since I-U is strictly increasing $((I-U)'(v)=1-U'(v)>0)$.
However, if $U'(v) > 1$ for at least one v, then there is at least one inflection point v_i with $U'(v_i) > 1$ (the inflection points of U are the extrema of U'). Then there is a value $e=\tilde{e}$ such that the line $w=\bar{v}-\tilde{e}$ intersects the curve $w=U(\bar{v})$ in the point $(v_i, U(v_i))$. The curve traverses the line from below. Therefore and since U is bounded there have to be at least two further intersection points of the line and the curve: equ.(4.6) has at least 3 solutions for $e=\tilde{e}$. Clearly this is also true for all values of e in a neighborhood of $\tilde{e}$.

Consider now the case n=3. If $U'(v_i) > 1$ for at least one i and if moreover the inequalities $U'(v_1) > 1$, $U'(v_2) < 1$, $U'(v_3) > 1$ are not satisfied simultaneously then there exist two values

$e_o < e_1$ with the property: for $e < e_o$ and for $e > e_1$ equ. (4.6) has exactly one solution, and for $e_o < e < e_1$ it has exactly three solutions. This means there is a single hysteresis loop. However, if the condition $U'(v_1) > 1$, $U'(v_2) < 1$, $U'(v_3) > 1$ holds then there are two loops which are either separated (fig. 4.2a) or overlapping (fig. 4.2 b).

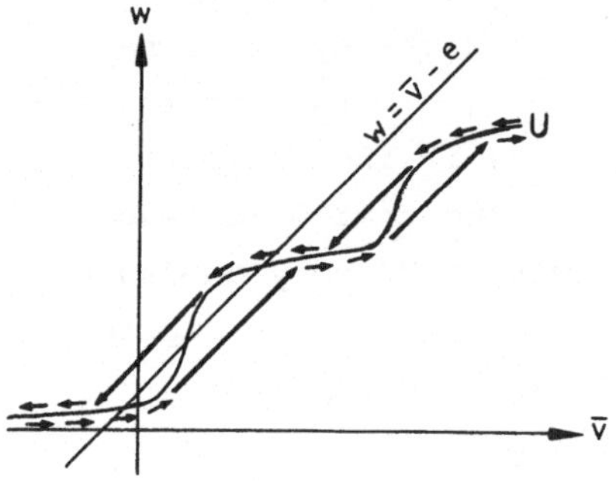

Fig. 4.2a

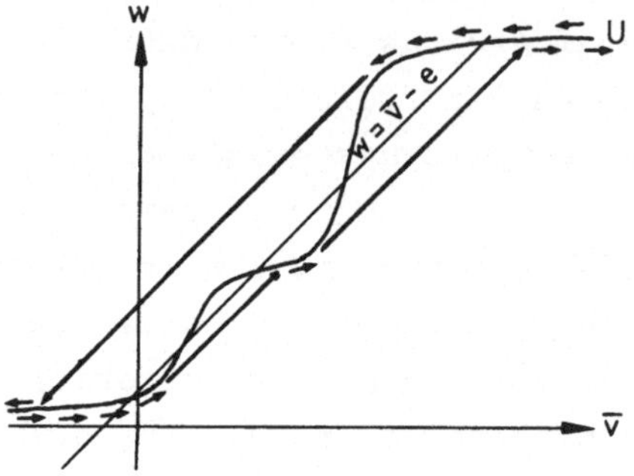

Fig. 4.2b

These two situations can be distinguished by the following criterion. There is a unique value $\bar{v} < v_1$ with $U'(\bar{v}) = 1$. Let α be the value where the line through the point $(\bar{v}, U(\bar{v}))$, which is tangent to $w = U(\bar{v})$ intersects the $\bar{v}$-axis. Let β be the number on the $\bar{v}$-axis determined by the tangent to the graph of U through the point $(\bar{\bar{v}}, U(\bar{\bar{v}}))$, $\bar{\bar{v}}$ denoting the value $\bar{v} > v_2$ satisfying $U'(\bar{v}) = 1$.

If $\alpha \leq \beta$ then there are two separated hysteresis loops and at most three steady states for a given value of e.

If $\alpha > \beta$ then there is an interval E such that for each $e \in E$ there are five solutions of equ. (4.6). In this case if there are two hysteresis loops these are overlapping.

The previous considerations show that hysteresis phenomena suggested by many authors (e.g. Cragg and Temperly, 1955) as models for short-term memory can occur with single neurons (or with single species populations in the sense of sect. 2.2).

Multiple hysteresis in neural interactions seems to have been discussed first by Wilson and Cowan, 1972. They considered two excitatory-inhibitory coupled cell types, but self-excitation was also necessary for the occurence of several steady states. For an example with arbitrarily many steady states in a homogeneous network see sect. 4.4.

4.2b) Pairs of neurons

According to the equs. (4.3) the stationary potentials $\bar{v}_1, \bar{v}_2$ of a system consisting of two neurons with external inputs e_1 and e_2 respectively satisfy

$$\bar{v}_1 = e_1 + H_{11} U_{11}(\bar{v}_1) + H_{12} U_{12}(\bar{v}_2) \tag{4.7a}$$

$$\bar{v}_2 = e_2 + H_{21} U_{21}(\bar{v}_1) + H_{22} U_{22}(\bar{v}_2) \; . \tag{4.7b}$$

If the functions U_{ij} were all linear then (4.7) would represent a system of linear equations. Then in general (i.e. whenever the determinant of the matrix of coefficients is different from zero) to every input (e_1, e_2) there corresponds exactly one solution $(\bar{v}_1, \bar{v}_2)$ of system (4.7). However, for non-linear functions U_{ij} (as it is always the case with neurons if only because of their thresholds and the non-negativity of their impulse frequencies) the situation is much more complicated.

In the following it is assumed that the functions $U_{ij}: \mathbb{R} \to \mathbb{R}_+$ are monotone increasing and bounded and that $H_{ij} \in \{-1, 1\}$ (the case $H_{ij}=0$ is included by the possibility $U_{ij}(v)=0$ for all v). Moreover self-excitation (i.e. $H_{11}=1$ or $H_{22}=1$) will be excluded from the consideration. Then essentially three types of coupling between the two neurons are left:

<u>Type 1</u> (reciprocal inhibition): $H_{12}=H_{21}=-1$

<u>Type 2</u> (excitatory-inhibitory coupling): $H_{12}=-1$, $H_{21}=1$

<u>Type 3</u> (reciprocal excitation): $H_{12}=H_{21}=1$.

Introducing the functions

$$\eta_1(\bar{v}_1) = [\bar{v}_1 + U_{11}(\bar{v}_1)]\ /H_{12} \text{ and } \eta_2(\bar{v}_2) = [\bar{v}_2 + U_{22}(\bar{v}_2)]\ /H_{21}$$

the system (4.7) can be written

$$\eta_1\ (\bar{v}_1) = U_{12}(\bar{v}_2) + \tilde{e}_1 \quad (4.8a)$$

$$\eta_2\ (\bar{v}_2) = U_{21}(\bar{v}_1) + \tilde{e}_2\ , \quad (4.8b)$$

$\tilde{e}_i = e_i/H_{ij}$.

The functions η_1 and η_2 are strictly monotone. Therefore with $f(\bar{v}_2) = \eta_1^{-1}\ (U_{12}(\bar{v}_2) + \tilde{e}_1)$ and $g(\bar{v}_1) = \eta_2^{-1}(U_{21}(\bar{v}_1)\ + \tilde{e}_2)$ the system (4.8) is equivalent to

$$\bar{v}_1 = f(\bar{v}_2)$$

$$\bar{v}_2 = g(\bar{v}_1)\quad .$$

Inserting the second equation into the first one reduces the problem to that of the single equation

$$\bar{v}_1 = f\circ g\ (\bar{v}_1)\quad . \quad (4.9)$$

In case of $H_{12} = -1$, $H_{21} = 1$ the function $f \circ g$ is monotone decreasing. Therefore the first result is:

Lemma 4.2. For a pair of excitatory-inhibitory coupled neurons (type 2) there exists one and only one steady state for each given constant input, provided there is no self-excitation.

Both for type 1 and for type 2 the function $f \circ g$ is monotone increasing. The situation is very similar to that of equ. (4.6), and the whole discussion there on hysteresis phenomena carries over to equ. (4.9). In particular it can be shown that

Lemma 4.3. Let the functions U_{ij}, $i,j=1,2$, be differentiable and assume there are numbers v_1^*, v_2^* such that

$$a = U_{12}'(v_2^*) = \max_{v\in\mathbb{R}} U_{12}'(v)\quad ,\ b = U_{21}'(v_1^*) = \max_{v\in\mathbb{R}} U_{21}'(v)\ .$$

Let $H_{12} \cdot H_{21}=1$ (i.e. either mutual inhibition or mutual excitation is given).

Then

(i) if $a \cdot b<1$ then to each pair (e_1,e_2) there corresponds a unique steady state $(\bar{v}_1,\bar{v}_2)$, satisfying equs. (4.8),

(ii) if $a \cdot b>1$ and $U_{11}=U_{22}=0$ (i.e. there is no self-inhibition) then there are stimulus configurations (e_1,e_2) giving at least three different steady states.

Proof. Let $H_{12}=H_{21}=1$ (the proof for $H_{12}=H_{21}=-1$ proceeds analogously).

$$(f\circ g)'(v_1)=(\eta_1^{-1})'(w_1)\cdot U_{12}'(g(v_1)) \quad\cdot\quad (\eta_2^{-1})'(w_2)\cdot U_{21}'(v_1), \qquad (4.10)$$

where $\qquad w_1=U_{12}(g(v_1)+e_1), \quad w_2=U_{21}(v_1)+e_2$.

Since $(\eta_i^{-1})'(w)\leq 1$ for $i=1,2$ and for all $w \in \mathbb{R}$ the inequality

$$(f\circ g)'(v_1) \leq a\cdot b$$

holds.

Therefore if $a \cdot b<1$ equ. (4.9) has exactly one solution which proves part (i).

If $U_{11}=U_{22}=0$ then $(\eta_i^{-1})'(w)=1$ which simplifies extremely the expression (4.10). There is some value $v_1=v_1^*$ with $U_{21}'(v_1^*)=b$ and some value $e_2=e_2^*$ with $U_{12}'(U_{21}(v_1^*)+e_2^*)=a$. Then $(f\circ g)'(v_1)=a\cdot b$ independent of e_1. Choose a value $e_1=e_1^*$ such that $v_1^*=U_{12}(g(v_1^*))+e_1^*= =f\circ g(v_1^*)$.

If $1<a\cdot b=(f\circ g)'(v_1^*)$ then, because of the boundedness of $f\circ g$, besides v_1^* there are two further solutions of equ. (4.9) for $e_1=e_1^*$ and $e_2=e_2^*$, Q.E.D.

Due to the complexity of the expression (4.10) the conditions for multiple steady states in case of non-vanishing self-inhibition are cumbersome. However, it can be concluded from equ. (4.10) that self-inhibition leads to smaller values of $(f\circ g)'$.

Therefore self-inhibition tends to decrease the number of steady states.
It may happen that the steady state is unique even if $a \cdot b > 1$. The domain of inputs (e_1, e_2) allowing several steady states is reduced by self-inhibition.

4.2. c) Arbitrarily many neurons

In this subsection conditions are derived that, given a constant input vector $e=(e_1, e_2, \ldots, e_n)$, the system (4.3) of nonlinear equations for the vector $\bar{v}=(\bar{v}_1, \bar{v}_2, \ldots, \bar{v}_n)$ of stationary membrane potentials has exactly one solution $\bar{v}=\bar{v}(e)$. It is assumed that $H_{ij} \in \{-1, 0, 1\}$, that $H_{ii} = -1$ and that the functions $U_{ij}: \mathbb{R} \longrightarrow \mathbb{R}_+$ are nonnegative and monotone increasing (but not necessarily bounded), $i,j=1,2,\ldots,n$. Since $\eta_i(\bar{v}_i)=\bar{v}_i+U_{ii}(\bar{v}_i)$ is a strictly increasing function, the system (4.3) can be written in the form

$$\bar{v}_i = \eta_i^{-1}\left(e_i + \sum_{j \neq i} H_{ij} U_{ij}(\bar{v}_j)\right), \quad i=1,2,\ldots,n. \tag{4.11}$$

Considering $\bar{v}_j$, $j=1,2,\ldots,n$, as variables the right hand side of equ. (4.11) can only take on values in the interval

$$I_i^e = \left\{\bar{v}_i \in \mathbb{R}: \eta_i^{-1}\left(e_i - \sum_{j \neq i} \underline{H}_{ij} \bar{U}_{ij}\right) \leq \bar{v}_i \leq \eta_i^{-1}\left(e_i + \sum_{j \neq i} \bar{H}_{ij} \bar{U}_{ij}\right)\right\} \tag{4.12}$$

where $\bar{U}_{ij} = \sup_{w \in \mathbb{R}} U_{ij}(w)$, $\underline{H}_{ij}=1$ if $H_{ij}=-1$, $\underline{H}_{ij}=0$ otherwise,

$\bar{H}_{ij}=1$ if $H_{ij}=1$, $\bar{H}_{ij}=0$ otherwise. Every solution $\bar{v}$ of (4.11) is an element of $\Omega^e = I_1^e \times I_2^e \times \ldots \times I_n^e$.

For the following theorem a Lipschitz condition is formulated:

<u>Condition 4.4.</u> a) For each function U_{ij} with $i \neq j$ there is a constant $L_{ij}^e \geq 0$ such that

$$|U_{ij}(w) - U_{ij}(\tilde{w})| \leq L_{ij}^e |w - \tilde{w}| \qquad \text{for all } w, \tilde{w} \in I_j^e,$$

b) for each function U_{ii} there is a constant $L_{ii}^{e} \geq 0$ such that

$$|U_{ii}(w)-U_{ii}(\tilde{w})| \geq L_{ii}^{e} |w-\tilde{w}| \quad \text{for all} \quad w,\tilde{w} \in I_i^e .$$

For a quadratic matrix M the <u>spectral radius</u> $\rho(M)$ is defined as the maximal modulus of all its eigenvalues:

$$\rho(M)= \max \{|\lambda|: \lambda \text{ eigenvalue of } M\} \quad .$$

Let the matrix $M^e=(m_{ij}^e)_{i,j=1,2,\dots,n}$ be defined by

$$m_{ii}^e = 0, \qquad m_{ij}^e = L_{ij}^e/(1+L_{ii}^e) \quad \text{for } i\neq j \quad . \qquad (4.13)$$

<u>Theorem 4.5.</u> If condition 4.4 is satisfied and if the spectral radius of the matrix M^e obeys the inequality $\rho(M^e)<1$ then the system of equations (4.3) has exactly one solution.

<u>Proof.</u> By $F_i(\bar{v})=\eta_i^{-1}(e_i+\sum_{i\neq j} H_{ij}U_{ij}(\bar{v}_j))$ a map $F: \Omega^e \to \Omega^e$ is defined.
For $w,\tilde{w} \in I_i^e$

$$(1+L_{ii}^e)|w-\tilde{w}| \leq |w-\tilde{w}|+ |U_{ii}(w)-U_{ii}(\tilde{w})|$$

$$=|w+U_{ii}(w)-\tilde{w}+U_{ii}(\tilde{w})| = |\eta_i(w)-\eta_i(\tilde{w})| \quad ,$$

consequently

$$|\eta_i^{-1}(r)-\eta_i^{-1}(\tilde{r})| \leq (1+L_{ii}^e)^{-1} |r-\tilde{r}| \qquad \text{for all } r,\tilde{r} \in \eta_i(I_i^e) \quad .$$

Hence it follows with condition 4.4a) that for all $v,\tilde{v} \in \Omega^e$

$$|F_i(v)-F_i(\tilde{v})| \leq (1+L_{ii}^e)^{-1} |\sum_{j\neq i} H_{ij}(U_{ij}(v_j)-U_{ij}(\tilde{v}_j))| \leq$$

$$\leq (1+L_{ii}^e)^{-1} \sum_{j\neq i} L_{ij}^e |v_j-\tilde{v}_j| \quad .$$

With the abbreviation $|x| = (|x_1|, |x_2|, \ldots, |x_n|)$ for a vector $x=(x_1,x_2,\ldots,x_n)$ this means

$$|F(v)-F(\bar{v})| \leq M^e\, |v-\tilde{v}| \qquad \text{for } v,\tilde{v} \in \Omega^e \ .$$

Since $\rho(M^e)<1$ the map F is a P-contraction on Ω^e, and it follows (see Ortega-Rheinboldt 1970, theorem 13.1.2) that F has one and only one fixed point. Q.E.D.

Remark 4.6. The theorem cited above also says if the requirements of a P-contraction are met (as it is the case with theorem 4.5), then for any $v^{(0)} \in \Omega^e$ the sequence $(v^{(k)})$ recursively defined by

$$v^{(k+1)} = F\, v^{(k)} \qquad , \ k=0,1,2,\ldots$$

converges toward the unique solution $\bar{v}$ of system (4.3). In this way the steady state $\bar{v}$ can be approximately computed. The error is estimated by

$$|v^{(k)}-\bar{v}| \leq (I-M^e)^{-1}\, M^e\, |v^{(k)}-v^{(k-1)}| \quad , \ k=1,2,\ldots$$

(I=identity matrix).

Remark 4.7. The spectral radius $\rho(M)$ of a matrix $M=(m_{ij})_{i,j=1,2,\ldots,n}$ is bounded as

(i) $$\rho(M) \leq \max_{1\leq i\leq n} \sum_{j=1}^{n} |m_{ij}| \ .$$

More precisely the following estimate holds:

(ii) $$\rho(M) \leq \rho(|M|) \quad ,$$

where $|M|$ denotes the matrix with entries $|m_{ij}|$, $i,j=1,2,\ldots,n$. If M is nonnegative (i.e., $m_{ij} \geq 0$) and if $(x_1,x_2,\ldots,x_n)$ is a positive vector (i.e. $x_i > 0$) then

(iii) $$\min_{1\leq i\leq n} \frac{1}{x_i} \sum_{j=1}^{n} m_{ij}x_j \leq \rho(M) \leq \max_{1\leq i\leq n} \frac{1}{x_i} \sum_{j=1}^{n} m_{ij}x_j$$

(see Varga, 1962, chap.2.3, exercise 2).

Remark 4.8. It is evident from the definition (4.13) of the matrix M^e that for sufficiently strong self-inhibition in the network the spectral radius $\rho(M^e) < 1$ and thus the steady state is uniquely determined.

Remark 4.9. The optimal choice of the Lipschitz constants in condition 4.4 for functions U_{ij} which are differentiable in the interval I_i^e is given by

$$L_{ij}^e = \sup_{w \in I_i^e} U'_{ij}(w) \qquad (i \neq j),$$

$$L_{ii}^e = \inf_{w \in I_i^e} U'_{ii}(w) \quad .$$

Example 4.10. The theorem (4.5) will be applied to networks with lateral inhibition realized in the retina of Limulus and to the analogous networks with lateral excitation. The former are described by the system (2.2) or, equivalently, by the system (4.3) with the specializations $H_{ij} = -1$ and

$$U_{ij}(w) = K_{ij}\, m(b_j m(w) - r_{ij}) = K_{ij}\, m(b_j w - r_{ij}),\ i,j=1,2,\ldots,n. \tag{4.14}$$

It follows from

$$|m(w) - m(\tilde{w})| \leq |w - \tilde{w}| \tag{4.15}$$

that

$$|U_{ij}(w) - U_{ij}(\tilde{w})| \leq b_j K_{ij} |w - \tilde{w}| \quad .$$

Therefore

$$L_{ij}^e = b_j K_{ij} \ (i \neq j) \text{ and } L_{ii}^e = 0 \tag{4.16}$$

are suitable constants for condition 4.4. Theorem 4.5 gives

Corollary 4.11. If the spectral radius of the matrix $(b_{ij})_{i,j=1,\ldots,n}$ defined by $b_{ij} = b_j K_{ij} (i \neq j)$, $b_{ii} = 0$ is smaller than unity, then the Hartline-Ratliff system (2.2) has exactly one solution.

This result generalizes theorem 2 of (Hadeler, 1974), where additionally $r_{ij} = 0$ and $K_{ii} = 0$ were assumed.

A special system with lateral excitation is given by the equs.(4.3) if $H_{ij}=1$ for $i\neq j$, $H_{ii}=-1$, $b_j>0$, $r_{ij}\geq 0$, and U_{ij} just as in (4.14). The constants L^e_{ij} can be chosen as in (4.16).
Therefore the criterion for lateral excitation equals that for lateral inhibition: If $\rho(B)<1$ then to each input $e=(e_1,e_2,\dots,e_n)$ there corresponds one and only one steady state $\bar{v}=(\bar{v}_1,\bar{v}_2,\dots,\bar{v}_n)$ of the network.

The condition of theorem 4.5 is in a certain sense optimal: In the situation of lemma 4.3 it follows from remark 4.9 that $\rho(M^e)=ab$, and therefore $\rho(M^e)<1$ is necessary and sufficient for the uniqueness of the steady state. However, in general as demonstrated by the examples in lemma 4.2 and in theorem 4.15 below (see also remark 4.16), the steady state can be unique even if the spectral radius is arbitrarily large. This unsatisfactory state of affairs can be improved somewhat by the following condition.

<u>Theorem 4.12.</u> (Ostrowski, 1966, th.24.2). Let $D\subset\mathbb{R}^n$ be a convex set, and let $F=(f_1,f_2,\dots,f_n):D\to\mathbb{R}^n$ be a map with continuously differentiable functions $f_i:D\to\mathbb{R}$, $i=1,2,\dots,n$.
If the determinant

$$K(\xi^{(1)},\xi^{(2)},\dots,\xi^{(n)})=\begin{vmatrix}\frac{\partial f_1}{\partial v_1}(\xi^{(1)}), & \frac{\partial f_1}{\partial v_2}(\xi^{(1)}), & \dots, & \frac{\partial f_1}{\partial v_n}(\xi^{(1)})\\ \vdots & \vdots & & \vdots\\ \frac{\partial f_n}{\partial v_1}(\xi^{(n)}), & \frac{\partial f_n}{\partial v_2}(\xi^{(n)}), & \dots, & \frac{\partial f_n}{\partial v_n}(\xi^{(n)})\end{vmatrix}\neq 0 \quad (4.17)$$

for arbitrary vectors $\xi^{(1)},\xi^{(2)},\dots,\xi^{(n)}\in D$, then the equation $F(\xi)=0$ has at most one solution in D.—

In order to apply theorem 4.12 to system (4.3) define

$$f_i(v) = f_i(v_1, v_2, \dots, v_n) = e_i - v_i + \sum_{j=1}^{n} H_{ij} U_{ij}(v_j) \tag{4.18}$$

and assume $f_i : D \longrightarrow \mathbb{R}$ to be continuously differentiable on a suitable convex set D, e.g. $D = I_1^e \times I_2^e \dots \times I_n^e$ (see equ.(4.12)). Then

$$\frac{\partial f_i}{\partial v_j} (\xi^{(i)}) = H_{ij} U'_{ij} (\xi_j^{(i)}) = H_{ij} \tilde{\xi}_j^{(i)} \qquad (i \neq j), \tag{4.19}$$

$$\frac{\partial f_i}{\partial v_i} (\xi^{(i)}) = -1 - U'_{ii}(\xi_i^{(i)}) = - \tilde{\xi}_i^{(i)} \tag{4.20}$$

with nonnegative numbers $\tilde{\xi}_j^{(i)}$, $\tilde{\xi}_i^{(i)} > 1$.

For n=2

$$K(\xi^{(1)}, \xi^{(2)}) = \tilde{\xi}_1^{(1)} \tilde{\xi}_2^{(2)} - H_{12} H_{21} \tilde{\xi}_2^{(1)} \tilde{\xi}_1^{(2)} .$$

The last expression is positive if H_{12} and H_{21} have different signs(type 2 in sect. 4.2.b) or if in case of equal signs (types 1 and 3) the inequality

$$\overline{U}'_{12} \cdot \overline{U}'_{21} < (1 + \underline{U}'_{11})(1 + \underline{U}'_{22})$$

holds, where

$$\overline{U}'_{ij} = \sup_{v \in \mathbb{R}} U'_{ij}(v), \qquad \underline{U}'_{ij} = \inf_{v \in \mathbb{R}} U'_{ij}(v)$$

Theorem 4.12 says that under these conditions there is at most one steady state.
The existence of a steady state can be concluded from theorem 4.1.
These results coincide with the observations in lemma 4.2 and generalize lemma 4.3 demonstrating the effectiveness of theorem 4.12.

Recalling that in the case of a single neuron with self-inhibition and of a pair of neurons with excitatory-inhibitory coupling the steady state is always unique for arbitrary functions U_{ij}

and input e it becomes a natural question how to characterize those neural networks for which this property is a consequence of the coupling type alone. In such networks hysteresis phenomena are excluded and the state of the network is uniquely related to its (constant) input.

A coupling type is determined by a matrix $H=(H_{ij})_{i,j=1,2,\dots,n}$ with $H_{ij}=-1,0,$ or 1 according as neuron j acts inhibitory, not at all, or excitatory on neuron i respectively. The desired characterisation has to be expressed in form of conditions on the matrix H.

<u>Theorem 4.13.</u> Let $H=(H_{ij})_{i,j=1,2,\dots,n}$ be a matrix with $H_{ij}\in\{-1,0,1\}$. Then the following two conditions are equivalent:

(i) For each vector $e=(e_1,e_2,\dots,e_n)$ and each set of continuously differentiable, nonnegative, monotone increasing, and bounded functions U_{ij}, $i,j=1,2,\dots,n$, there is exactly one solution $\bar{v}=(\bar{v}_1,\bar{v}_2,\dots,\bar{v}_n)$ of system (4.3).

(ii) For each $p\in\{1,2,\dots,n\}$ and each ordered sequence $(i_1,i_2,\dots,i_p)$ of pairwise different indices $i_1, i_2,\dots,i_p\in\{1,2,\dots,n\}$ the inequality

$$H_{i_1i_2}\cdot H_{i_2i_3}\cdot\ \dots\ \cdot H_{i_{p-1}i_p}\cdot H_{i_pi_1}\neq 1$$

holds.

The proof of this theorem is based on

<u>Lemma 4.14.</u> Let $A=(a_{ij})$ be an $n\times n$-matrix with the properties

(i) $a_{ii}<0$, $i=1,2,\dots,n$,

(ii) for each $p\in\{1,2,\dots,n\}$ and each system $i_1,i_2,\dots,i_p\in\{1,2,\dots,n\}$ of pairwise different indices the inequality

$$a_{i_1i_2}\cdot a_{i_2i_3}\cdot\ \dots\ \cdot a_{i_pi_1}\leq 0 \tag{4.21}$$

holds.

Then the determinant of A satisfies the inequality

$$|A|>0,\quad \text{if } n \text{ is even},$$
$$|A|<0,\quad \text{if } n \text{ is odd}.$$

Proof. $|A|$ has the expansion

$$|A| = \sum_{\nu_1,\nu_2,\dots,\nu_n} \operatorname{sign}(\nu_1,\nu_2,\dots,\nu_n)\ a_{1\nu_1} a_{2\nu_2} \cdots a_{n\nu_n},$$

where the summation runs through all permutations $(\nu_1,\nu_2,\dots,\nu_n)$ of the numbers $1,2,\dots,n$. Each ordered sequence $\{i_1,i_2,\dots,i_p\} \subset \{1,2,\dots,n\}$ defines a cyclic permutation of length p with sign $(-1)^{p+1}$. Each permutation $(\nu_1,\nu_2,\dots,\nu_n)$ can be represented as a product of m cycles with lengths $p_1,p_2,\dots,p_m$, and

$$\operatorname{sign}(\nu_1,\nu_2,\dots,\nu_n) = \prod_{k=1}^{m} (-1)^{p_k+1} = (-1)^m \cdot (-1)^n . \tag{4.22}$$

According to this cycle representation of $(\nu_1,\nu_2,\dots,\nu_n)$ the product $a_{1\nu_1} a_{2\nu_2} \cdots a_{n\nu_n}$ can be written as the product of m nonpositive factors of the form (4.21). Therefore, if

$a_{1\nu_1} \cdot a_{2\nu_2} \cdot \dots \cdot a_{n\nu_n} \neq 0$,

$$\operatorname{sign}[\operatorname{sign}(\nu_1,\nu_2,\dots,\nu_n) a_{1\nu_1} \cdot a_{2\nu_2} \cdot \dots \cdot a_{n\nu_n}] = (-1)^m (-1)^n (-1)^m = (-1)^n .$$

This together with condition (i) proves the lemma.

Proof of theorem 4.13. Condition (i) follows from condition (ii): In view of theorem 4.1 system (4.3) has a solution. The solution is uniquely determined as can be seen by help of theorem 4.12 and the equs. (4.19) and (4.20): Define

$$a_{ij} = H_{ij} U'_{ij} (\xi_j^{(i)}) \quad (i \neq j), \ a_{ii} = -1 - U'_{ii} (\xi_i^{(i)}) .$$

Then lemma 4.14 shows that the inequality (4.17) of theorem 4.12 is always satisfied, which proves (i).
That condition (ii) is a consequence of condition (i) will be proved indirectly: Assume there is a number $p \in \{1,2,\dots,n\}$ and a set $\{i_1,i_2,\dots,i_p\} \subset \{1,2,\dots,n\}$ such that

$$H_{i_1 i_2} \cdot H_{i_2 i_3} \cdot \ldots \cdot H_{i_{p-1} i_p} \cdot H_{i_p i_1} = 1 \quad . \tag{4.23}$$

By renumeration $(i_1, i_2, \ldots, i_p) = (1, 2, \ldots, p)$ can be achieved. Moreover for $i=1,2,\ldots,p-1$ monotone increasing, bounded, and continuously differentiable functions $U_{i,i+1}$ and U_{p1} can be chosen such that the equation

$$\overline{v}_1 = H_{12}\, U_{12} \circ H_{23}\, U_{23} \circ \ldots \circ H_{p1}\, U_{p1}(\overline{v}_1) \tag{4.24}$$

has at least three solutions $\overline{v}_1^{(1)} < \overline{v}_1^{(2)} < \overline{v}_1^{(3)}$ (since the composed function on the right hand side is monotone increasing because of (4.23)). For all other functions U_{ij} assume $U_{ij}(\overline{v}_j)=0$ for $\overline{v}_j \leq \max\,(0, \overline{v}_1^{(3)})$. Then system (4.3) with $e=0$ is reduced to

$$\overline{v}_i = H_{i,i+1}\, U_{i,i+1}(\overline{v}_{i+1}), \qquad i=1,2,\ldots,p-1$$

$$\overline{v}_p = H_{p1}\, U_{p1}(\overline{v}_1)$$

in the domain $D = \{ (\overline{v}_1, \overline{v}_2, \ldots, \overline{v}_n) : \overline{v}_i \leq \max\,(0, \overline{v}_1^{(3)}) \}$.

This system can be resolved to the variable $\overline{v}_1$ in the form (4.24), and hence it has at least three different solutions which is a contradiction to condition (i). Q.E.D.

The condition (ii) of theorem 4.13 simply says there are no positive feedback loops within the network. One conclusion from theorem 4.13 is that hysteresis phenomena can only occur in networks containing at least one positive feedback (an extreme case is $H_{ii}=1$ for some i). Positive feedback can be generated by way of two inhibitory couplings (e.g. $H_{12}=H_{21}=-1$ results in $H_{12} \cdot H_{21} = +1$). An example of a three cell network satisfying the condition (ii) of theorem 4.13 is depicted in the diagram of fig. 4.4.

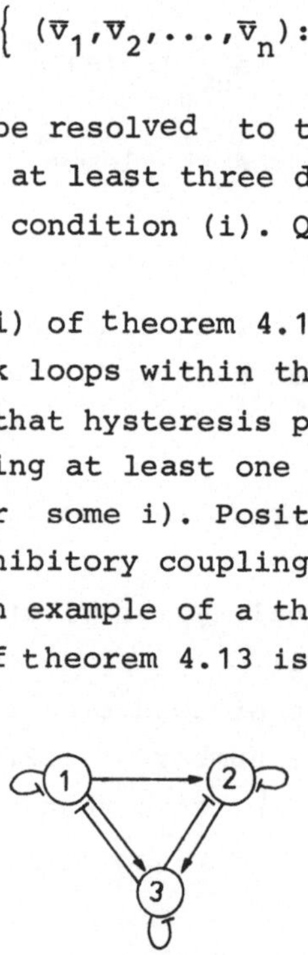

Fig. 4.4

$(H_{ii} = H_{13} = H_{23} = -1,\ H_{21} = H_{32} = H_{31} = 1)$.

4.3. Input-output behavior of stationary networks

According to the previous investigations, generally for each temporally stationary input $e=(e_1,\dots,e_n)$ to the network there corresponds a steady state $\bar{v}=\bar{v}(e)=(\bar{v}_1(e),\bar{v}_2(e),\dots,\bar{v}_n(e))$ satisfying the system of equations (4.3). In this subsection $\bar{v}(e)$ is considered as output of the network, and the question is in which way $\bar{v}$ depends on e. If $\bar{v}(e)$ is uniquely determined this question concerns the properties of the map

$$e \longmapsto \bar{v}(e) \quad .$$

If $\bar{v}(e)$ is not unique then, since the system can be only in one of the steady states, the problem is through which of the several steady states the system will pass if e is continuously changed (in such a way that the system is always in a steady state), or more precisely, are there continuous branches of the multi-valued map $e \longmapsto \{\bar{v}(e)\}$, and in which way does $\bar{v}(e)$ depend on e if this map is restricted to one of its branches?

All these questions can be genericly answered by the implicit function theorem (contained in every basic book on calculus):

For its application the functions U_{ij} are supposed to be continuously differentiable. It is assumed that for the fixed input $e=e^*$ the system is in the steady state v^* . Then following the implicit function theorem there is a neighborhood $N(e^*)\subset \mathbb{R}^n$ of e^* such that for all inputs $e\in N(e^*)$ the system of equations (4.3) has a uniquely determined solution $\tilde{v}(e)$ with the property that $\tilde{v}:N(e^*)\longrightarrow \mathbb{R}^n$ is a continuously differentiable function satisfying $\tilde{v}(e^*)=v^*$, if

$$\Delta(v^*)=\det\left[\left(H_{ij}\frac{dU_{ij}}{dv_j}(v_j^*)-\delta_{ij}\right)_{i,j=1,2,\dots,n}\right] \neq 0 \, , \tag{4.25}$$

where $\delta_{ij}=0$ $(i\neq j)$, $\delta_{ii}=1$.

The essential conclusion from this proposition is that a bifurcation of one steady state into several steady states can only occur at a state v^* where the determinant $\Delta(v^*)$ vanishes. In such a state it may happen that an arbitrarily small perturbation of the input e^* to another input $e^* + \varepsilon$ causes the system to make a discontinuous jump from v^* to a state $\bar{v}(e^*+\varepsilon)$ with $\| v^* - \bar{v}(e^* + \varepsilon)\| > \alpha > 0$, α independent of the smallness of ε (a catastrophe in the language of catastrophe theory).
However, if $\Delta(v^*) \neq 0$ then a small continuous change of e away from e^* will cause the system to change continuously according to the function $\tilde{v}(e)$ guaranteed by the implicit function theorem. Some insight into this change is possible by the following consideration: If $\Delta(v^*) \neq 0$ and the functions U_{ij} are continuously differentiable then the differentiation of the equs. (4.3) with respect to e results in

$$I + \left[\left(H_{ij} \frac{dU_{ij}}{dv_j} (\bar{v}_j(e))\right)_{i,j=1,\dots,n} - I\right] \frac{\partial \bar{v}}{\partial e}(e) = 0, \qquad (4.26)$$

and hence

$$\left(\frac{\partial \bar{v}_i}{\partial e_j}(e)\right)_{i,j=1,\dots,n} = \frac{\partial \bar{v}}{\partial e}(e) = \left[I - \left(H_{ij} \frac{dU_{ij}}{dv_j} (\bar{v}_j(e))\right)\right]^{-1}. \qquad (4.27)$$

This equation allows one to compute the alteration of the state of the network with respect to the input if the quantities H_{ij}, U_{ij} and the steady state $\bar{v}$ are known. On the other hand knowledge (e.g. by measurements) of the derivatives $\frac{\partial \bar{v}_i}{\partial e_j}$ may help to determine the interaction matrices (H_{ij}), (U_{ij}) by evaluating the equation

$$\left(H_{ij} \frac{dU_{ij}}{dv_j} (\bar{v}_j)\right)_{i,j=1,\dots,n} = I - \left[\frac{\partial \bar{v}}{\partial e}(e)\right]^{-1} \qquad (4.28)$$

which follows from equ. (4.26) on the assumption that the last inverse exists. Equ. (4.28) may be of some practical interest as it enables one (at least in principle), simply by slow variations of the external stimulus e, to determine the interaction in the network without destroying or isolating parts of the network. The crucial supposition of this method is that the stationary behavior of the net is governed by equations of the form (4.3).

The preceding considerations will be illustrated by examining the case n=2 in greater detail. Evaluation of equ.(4.27)results in

$$\begin{pmatrix} \frac{\partial \tilde{v}_1}{e_1} & \frac{\partial \tilde{v}_1}{e_2} \\ \frac{\partial \tilde{v}_2}{\partial e_1} & \frac{\partial \tilde{v}_2}{\partial e_2} \end{pmatrix} = \frac{1}{\Delta_1} \cdot \begin{pmatrix} 1-H_{22}U'_{22}(\tilde{v}_2), & H_{12}U'_{12}(\tilde{v}_2) \\ H_{21}U'_{21}(\tilde{v}_1), & 1-H_{11}U'_{11}(\tilde{v}_1) \end{pmatrix} \tag{4.29}$$

where $\Delta_1=(1-H_{11}U'_{11}(\tilde{v}_1))(1-H_{22}U'_{22}(\tilde{v}_2))-H_{12}H_{21}U'_{12}(\tilde{v}_2)U'_{21}(\tilde{v}_1)$.

Obviously the sign of Δ_1 is essential for the distinction whether $\tilde{v}_i$ will increase or decrease as a function of e_j (the other component of e being held fixed).

Equ.(4.28) results in (note $\tilde{v}=\tilde{v}(e)$)

$$\begin{pmatrix} H_{11}\ U'_{11}(\tilde{v}_i) & , & H_{12}\ U'_{12}(\tilde{v}_2) \\ H_{21}\ U'_{21}(\tilde{v}_1) & , & H_{22}\ U'_{22}(\tilde{v}_2) \end{pmatrix} = \begin{pmatrix} \Delta_2-\frac{\partial \tilde{v}_2}{\partial e_2}(e) & , & \frac{\partial \tilde{v}_1}{\partial e_2}(e) \\ \frac{\partial \tilde{v}_2}{\partial e_1}(e) & , & \Delta_2-\frac{\partial \tilde{v}_1}{\partial e_1}(e) \end{pmatrix} \cdot \frac{1}{\Delta_2} \tag{4.30}$$

with $\Delta_2 = \frac{\partial \tilde{v}_1}{\partial e_1}(e) \cdot \frac{\partial \tilde{v}_2}{\partial e_2}(e) - \frac{\partial \tilde{v}_1}{\partial e_2}(e) \cdot \frac{\partial \tilde{v}_2}{\partial e_1}(e)$.

The right hand side of equ.(4.30)is sometimes accessible experimentally. If boundary conditions for the function U_{ij} are known it is possible to derive an approximation of U_{ij} if sufficiently many values $U'_{ij}(\tilde{v}_j)$ are obtained from equ. (4.30).

4.4 An example of spatial hysteresis

This subsection contains a very simple example of a network with extremely many steady states. Consider the system (2.2) of lateral inhibition (which is a special case of (4.1) and hence of (4.3)). Assume a homogeneous structure of the network: all lateral inhibition coefficients K_{ij} are identical, $K_{ij}=K$, $(i\neq j)$ (this condition may apply approximately to a local neural aggregate), and the input array is constant, $e_i=\eta$, $i=1,2,\dots,n$. Moreover let self-inhibition be excluded ($K_{ii}=0$) and all the thresholds $r_{ij}=0$. Then with the transformation (4.4) the potential representation of the system is

$$\bar{v}_i=\eta-K\sum_{j\neq i} m(\bar{v}_j) \quad , \qquad i=1,2,\dots,n. \tag{4.31}$$

Theorem 4.15. (i) If $\eta\leq 0$ then system (4.31) has exactly one solution, given by $\bar{v}_i=\eta$, $i=1,2,\dots,n$.

(ii) If $\eta>0$ and $K<1$ then system (4.31) has exactly one solution, given by $\bar{v}_i=\eta/(1+(n-1)K)$, $i=1,2,\dots,n$.

(iii) If $\eta>0$ and $K>1$ then there are exactly 2^n-1 different solutions of system (4.31). These are just all vectors $(\bar{v}_1,\bar{v}_2,\dots,\bar{v}_n)$ satisfying

$$\bar{v}_{i_1}=\bar{v}_{i_2}=\dots=\bar{v}_{i_k}=\frac{1}{1+(k-1)K}\eta; \quad \bar{v}_i=\frac{1-K}{1+(k-1)K}\eta \tag{4.32}$$

for $i\neq i_1,\dots,i_k$, where k is an arbitrary integer with $1\leq k\leq n$, and $\{i_1,i_2,\dots,i_k\}$ an arbitrary subset of $\{1,2,\dots,n\}$.

Proof. (i) Equ. (4.31) implies $\bar{v}_i\leq\eta$; hence $m(\bar{v}_j)=0$. With (4.31) $\bar{v}_i=\eta$.

(ii) Let $v=(v_1,v_2,\dots,v_n)$ be any solution of (4.31). The right hand side of (4.31) only depends on the positive components $v_{i_1},v_{i_2},\dots,v_{i_k}$ of v. Addition of the equs. (4.31) with the indices $i_1,i_2,\dots,i_k$ gives

$$(1+ (k-1)K) \sum_{\varkappa=1}^{k} v_{i_\varkappa} = k\eta \quad . \tag{4.33}$$

Hence with equ. (4.31) for $i \neq i_1, i_2, \ldots, i_k$

$$v_i = \eta - \frac{kK}{1+(k-1)K}\eta = \frac{1-K}{1+(k-1)K}\eta \quad . \tag{4.34}$$

For $K < 1$ the last expression is positive. Therefore $k=n$, and all components of v are positive. In this case (4.31) is a linear system of equations, the coefficient matrix of which has a non-vanishing determinant if $K \neq 1$. This implies that there is at most one positive solution. It is easily verified that $v_i = \eta/(1+(n-1)K)$ is a solution.
(iii) Obviously the 2^n-1 vectors defined by (4.32) are solutions of equ. (4.31), if $\eta > 0$ and $K > 1$. Using the notations in the proof of (ii) the equations in (4.31) with the indices $i_1, i_2, \ldots, i_k$ form a linear system with a unique solution. This fact together with the relations (4.34) implies that there cannot be more than 2^n-1 solutions, Q.E.D.

Remark 4.16. Theorem 4.15(ii) gives an example of a system with lateral inhibition having a unique steady state although the spectral radius of the inhibition coefficient matrix exceeds 1 (compare corollary 4.11). By help of the criterion in remark 4.7 it is easily computed (with $x_i=1$) that the spectral radius of the matrix (b_{ij}) with $b_{ij}=K$ for $i \neq j$, $b_{ii}=0$ is $(n-1)K$.

With the condition of theorem 4.15(iii) the system has 2^n-1 steady states for a given input. In which of these states the system will actually be depends on the history and on the dynamics of the system, neither of which have been considered in this chapter.

5. Local stability analysis of nets with finitely many neurons

5.1 Introduction

Whereas the previous chapter dealt with the stationary behavior of neural networks this chapter is the first one concerned with the time-dependent properties, i.e. the dynamics of such nets. These properties are reflected by the time-dependent solutions of system (1.15), which for convenience is restated here:

$$v_i = v_{io} + \sum_{j=1}^{n} U_{ij}(v_j)*h_{ij} + \bar{S}_i(E_i)*\bar{h}_i \quad , \quad i=1,2,\ldots,n \quad . \tag{5.1}$$

It is difficult to investigate this system of Volterra integral equations for inputs $E_i = E_i(t)$ which are arbitrary functions of time. Therefore the first step in the analysis is restricted to constant inputs or to at most small temporal perturbations of constant inputs.

It is assumed throughout this chapter that $\bar{v}=\mathrm{col}(\bar{v}_1,\bar{v}_2,\ldots,\bar{v}_n)$ is any of the steady state solutions of system (5.1) belonging to some constant input $E=\bar{E}=\mathrm{col}(\bar{E}_1,\bar{E}_2,\ldots,\bar{E}_n)$, i.e. $\bar{v}=\bar{v}(\bar{E})$. For any other input $E(t)$ the perturbation input $e(t)$ with respect to $\bar{E}$ is defined as

$$e_i(t) = \bar{S}_i(E_i)*h_i - \bar{S}_i(\bar{E}_i)*h_i, \; i=1,2,\ldots,n.$$

Then the perturbation variables $w_i(t) = v_i(t)-\bar{v}_i$ with respect to $\bar{v}$ satisfy the system

$$w_i(t) = \sum_{j=1}^{n} [U_{ij}(w_j+\bar{v}_j)-U_{ij}(\bar{v}_j)]*h_{ij} + e_i(t), \tag{5.2}$$

which is equivalent to system (5.1). The solution $\bar{v}$ of (5.1) corresponds to the solution 0 of (5.2).
With

$$f_i(t)=\sum_{j=1}^{n}\int_{-\infty}^{0}[U_{ij}(v_j(t'))-U_{ij}(\bar{v}_j)]\; h_{ij}(t-t')dt' + e_i(t) \tag{5.3}$$

system (5.1) can be written

$$w_i(t)=f_i(t)+\sum_{j=1}^{n}\int_{0}^{t} [U_{ij}(v_j(t')-U_{ij}(\bar{v}_j)]\ h_{ij}(t-t')dt' \quad . \qquad (5.4)$$

It is the purpose of this chapter to investigate the behavior of solutions v(t) of system (5.1) near $\bar{v}$, i.e. of solutions w(t) of the perturbed system (5.4) in the neighborhood of zero. More precisely some notions and criteria of local stability of steady states shall be developed.

Definition 5.1. The steady state solution $\bar{v}=(\bar{v}_1,\ldots,\bar{v}_n)$ of system (5.1) is called (locally) stable if (and only if) to every $\varepsilon>0$ a number $\delta=\delta(\varepsilon)>0$ is associated such that for any continuous function $f: \mathbb{R}_+ \longrightarrow \mathbb{R}^n$ with $\|f(t)\| \leq \delta$, $t\in\mathbb{R}_+$, all solutions $w: \mathbb{R}_+ \longrightarrow \mathbb{R}^n$ of system (5.4) satisfy $\|w(t)\| \leq\varepsilon$ for all $t\in\mathbb{R}_+$.
If additionally there is a constant $\varkappa>0$ such that

$$\|f(t)\| \leq\varkappa,\ t\in\mathbb{R}_+, \qquad \lim_{t\to\infty} f(t) = 0$$

implies $\lim_{t\to\infty} w(t) = 0$ then (and only then) $\bar{v}$ is called (locally) asymptotically stable.

In the following section the problem of local stability in the nonlinear system (5.1) will be reduced to that of the associated linear system. Then it is possible to use the well known stability criteria of the linear theory in favor of the nonlinear system .

5.2 The linearization principle

Assume the functions U_{ij}, i,j=1,2,...,n, to be differentiable at the point $\bar{v}_j$, i.e. there is a representation

$$U_{ij}(\xi+\bar{v}_j)= U_{ij}(\bar{v}_j) + U'_{ij}(\bar{v}_j)\xi + o_{ij}(\xi)$$

with

$$\lim_{\xi \to 0} o_{ij}(\xi)/\xi = 0 \quad . \tag{5.5}$$

Then according to equ. (5.4) system (5.1) is equivalent to

$$w_i(t)=f_i(t)+\sum_{j=1}^{n} \int_0^t [U'_{ij}(\overline{v}_j)w_j(t')+o_{ij}(w_j(t')] \, h_{ij}(t-t')dt'. \tag{5.6}$$

Define a matrix function $g: \mathbb{R}_+ \to \mathbb{R}^{n^2}$ by

$$g_{ij}(t) = U'_{ij}(\overline{v}_j) \, h_{ij}(t) \quad . \tag{5.7}$$

Then the system of linear Volterra integral equations

$$z(t)=f(t) + \int_0^t g(t-t') \, z(t')dt', \tag{5.8}$$

$z(t) \in \mathbb{R}^n$, is called the <u>linearized system</u>, with respect to $\overline{v}$, associated with (5.1).

The <u>resolvent</u> R corresponding to a continuous matrix kernel g is the uniquely determined continuous solution $R=(R_{ij})_{i,j=1,2,\ldots,n}$ of the matrix equation

$$R(t)= - g(t) + \int_0^t g(t-t')R(t')dt' \quad . \tag{5.9}$$

As is well known (see e.g. Miller, 1971) the solution of the linear system (5.8) (with suitable continuity assumptions on f and g) can be expressed in terms of the resolvent and the function f :

$$z(t)=f(t)- \int_0^t R(t-t') \, f(t')dt' \quad . \tag{5.10}$$

The following theorem will specify that a simple condition on the resolvent of the linearized system is sufficient to guarantee the asymptotic stability of steady states of the nonlinear system.

<u>Theorem 5.2.</u> The constant solution $\bar{v}=(\bar{v}_1,\dots,\bar{v}_n)$ of system (5.1) is asymptotically stable if the conditions (i)-(iv) hold for all i,j:

(i) the functions $U_{ij}: \mathbb{R} \to \mathbb{R}$ and $h_{ij} : \mathbb{R}_+ \to \mathbb{R}$ are continuous, and

$$\int_0^\infty |h_{ij}(t)|\, dt < \infty \quad ,$$

(ii) the function U_{ij} is differentiable at the point $\bar{v}_j$,

(iii) the resolvent R associated with the kernel g of the with respect to $\bar{v}$ linearized system (5.8) satisfies

$$\int_0^\infty \| R(t) \|\ dt < \infty \quad ,$$

(iv) for each continuous $f: \mathbb{R}_+ \to \mathbb{R}$ equ. (5.6) has exactly one solution.

<u>Proof.</u> α) First it will be shown that $\bar{v}$ is stable.

Define

$$p_i(t,w) = \sum_{j=1}^{n} \int_0^t o_{ij}(w_j(t'))h_{ij}(t-t')dt', \quad t \geq 0 ,$$

$$p = \mathrm{col}\ (p_1,\dots,p_n) \ .$$

Then equ. (5.6) can be written

$$w(t)=f(t)+p(t,w)+\int_0^t g(t-t')w(t')dt' \quad . \tag{5.11}$$

It follows from equ. (5.10) that equ. (5.11) is equivalent to

$$w(t)=e(t)+p(t,w)-\int_0^t R(t-t')(e(t')+p(t',w))dt'. \tag{5.12}$$

Let $B=C_b(\mathbb{R}_+,\mathbb{R}^n)$ denote the Banach space consisting of all continuous and bounded functions $w: \mathbb{R}_+ \to \mathbb{R}^n$, with norm

$$\|w\|_B = \sup \left\{ \| w(t) \| \ : t \in \mathbb{R}_+ \right\} .$$

The assumptions on the functions h_{ij} imply that the resolvent R as a solution of equ. (5.9) exists and is continuous for $t \geq 0$.

Because of

$$\varkappa_1 = \sup \left\{ \int_0^t \|R(t-t')\| dt' : 0 \leq t < \infty \right\} < \infty$$

for any $y \in B$ the inequality

$$\left\| \int_0^t R(t-t')y(t')dt \right\| \leq \|y\|_B \int_0^t \|R(t-t')\| dt' \leq \varkappa_1 \|y\|_B ,$$

holds. Therefore the operator T defined by

$$Ty(t) = \int_0^t R(t-t')y(t')dt'$$

maps B into itself.

Because of (5.5), given $\rho > 0$ there is $\varepsilon > 0$ such that

$$\| o_{ij}(\xi) \| \leq \rho |\xi| \qquad \text{if } |\xi| \leq \varepsilon \tag{5.13}$$

independent of i,j. In particular

$$\| o_{ij}(\xi) \| \leq \rho\varepsilon \qquad \text{if } |\xi| \leq \varepsilon .$$

Let $w \in B$ and $\|w\|_B \leq \varepsilon$. Then

$$|p_i(t,w)| \leq \rho\varepsilon \sum_{j=1}^{n} \int_0^t |h_{ij}(t-t')| \, dt' .$$

Therefore $\|p(.,w)\|_B \leq \rho\varepsilon\varkappa_2$ with a constant $\varkappa_2$ depending only on the matrix of functions h_{ij}.

For $w \in B$, $\|w\|_B \leq \varepsilon$ and fixed $f \in B$ the operator $A: B \longrightarrow B$ defined by the right hand side of equ. (5.12),

$$Aw(t) = f(t) + p(t,w) - \int_0^t R(t-t') \, [e(t') + p(t',w)] \, dt' , \tag{5.14}$$

satisfies

$$\|Aw\|_B \leq (1+\varkappa_1)\ \|f\|_B + (\varkappa_1+\varkappa_2)\rho\varepsilon\ .$$

Choose $\rho>0$ such that $(\varkappa_1+\varkappa_2)\rho \leq \frac{1}{2}$. Then for each ε obeying the relation (5.13) the following condition holds:

$$\text{if } \|f\|_B \leq \delta \qquad \text{then} \quad \|Aw\|_B \leq \varepsilon \quad , \tag{5.15}$$

where $\delta = \varepsilon/(2(1+\varkappa_1))$.
The operator A maps the ball $D=\{w\in B: \|w\|_B \leq \varepsilon\}$ continuously into itself. The Schauder-Tychonoff fixed point theorem will be applied to A restricted to D, using the topology of uniform convergence on compact subsets of $\mathbb{R}_+$ in the space $C(\mathbb{R}_+, \mathbb{R}^n)$ of all continuous functions $\mathbb{R}_+ \longrightarrow \mathbb{R}^n$. Note that D is closed in $C(\mathbb{R}_+, \mathbb{R}^n)$ with respect to this topology. It remains to show that the set AD is relatively compact. According to the lemma of Arzela-Ascoli it suffices to prove that the functions in AD are uniformly bounded and equicontinuous on any compact subset $[0,T]\subset\mathbb{R}_+$, $T>0$. The first of these conditions is satisfied since $AD\subset D$.
For any two numbers t_1, t_2 with $0\leq t_1 < t_2 \leq T$ and any $w\in D$ the following estimates hold:

$$\begin{aligned}
\|Aw(t_1)-Aw(t_2)\| &\leq \|f(t_1)-f(t_2)\| + \|p(t_1,w)-p(t_2,w)\| \\
&\quad + \Big\|\int_0^{t_1} R(t_2-t')(f(t')+p(t',w))dt' - \int_0^{t_1} R(t_1-t')(f(t')+p(t',w))dt'\Big\| \\
&\leq \|f(t_1)-f(t_2)\| + \rho\varepsilon\varkappa_3\Big(\int_0^T \|h(t_2-t_1+s)-h(s)\|\,ds + \int_{t_1}^{t_2} \|h(t_2-t')\|\ dt'\Big) \\
&\quad + (\varepsilon/(2(1+\varkappa_1)) + \varkappa_2\rho\varepsilon)\Big(\int_0^T \|R(t_2-t_1+s)-R(s)\|\,ds + \int_{t_1}^{t_2} \|R(t_2-t')\|\,dt'\Big)\ ,
\end{aligned}$$

where the constant $\varkappa_3$ is independent of t_1, t_2 and w.
Since the functions h and R are uniformly continuous on compact intervals the estimates prove that AD is equicontinuous.

Therefore A has a fixed point in D. This fixed point is the solution of equ. (5.6), if f satisfies the first inequality in (5.15). Therefore ∇ is stable.

ß) In order to prove that ∇ is asymptotically stable assume $f \in B_o = \{ y \in B: \lim y(t)=0 \text{ as } t \to \infty \}$ and $\|f\|_B \leq \delta$, δ chosen as in part α) of this proof.
Then, according to part α) the solution w of equ. (5.6) satisfies $\|w\|_B \leq \varepsilon$ and w=Aw.
Consider the set $D_o = D \cap B_o$.
The operator A maps D_o into itself if for $w \in B_o$ each term on the right hand side of equ. (5.14) is an element of B_o.
It is not difficult to prove that for any two functions $g_1: \mathbb{R}_+ \to \mathbb{R}$ and $g_2: \mathbb{R}_+ \to \mathbb{R}$ the conditions g_1 continuous, $\lim g_1(t)=0$ as $t \to \infty$, and $\int_0^\infty |g_2(t)|\, dt < \infty$ imply

$$\lim_{t\to\infty} \int_0^t g_2(t-t')g_1(t')dt' = 0,$$

(see e.g. Corduneanu, theorem 2.7.2). Since o_{ij} is continuous and $\lim o_{ij}(\xi) = 0$ as $\xi \to 0$ and $w_j(t) \to 0$ as $t \to \infty$,

$$\lim_{t\to\infty} \int_0^t o_{ij}(w_j(t'))h_{ij}(t-t')dt' = 0 \quad .$$

Hence $p_i(t,w) \to 0$ as $t \to \infty$. Similarly the assumptions on R and f imply that $R * (f+p(.,w)) \in B_o$.

Just as in α) the Schauder-Tychonoff principle yields a fixed point of the operator A in the set D_1. This fixed point is the solution w of equ. (5.6) and satisfies $w(t) \to 0$ as $t \to \infty$. This remark completes the proof of the asymptotic stability of ∇. Q.E.D.

For the application of theorem 5.2 it is important to have a criterion on the resolvent R to satisfy condition (iii) of the theorem under the hypotheses of condition (i) (note

that g is defined via the functions h_{ij}).
Here a general theorem, due to Paley and Wiener, from the theory of linear Volterra integral equations is available, using Laplace transforms.

Theorem 5.3. (Paley and Wiener, 1934).

Suppose $g \in L^1(\mathbb{R}_+, \mathbb{R}^{n^2})$. Then the associated resolvent R is of class $L^1(\mathbb{R}_+, \mathbb{R}^{n^2})$ if and only if the determinant

$$\det(I- \int_0^\infty e^{-\lambda t} g(t)dt) \neq 0$$

for all complex numbers λ with $\mathrm{Re}\,\lambda \geq 0$.

The function

$$g^*(\lambda) = \int_0^\infty e^{-\lambda t} g(t)dt$$

is known as the Laplace transform of the kernel g.
With g as defined in (5.7) the equation

$$D(\lambda) = \det(I- g^*(\lambda)) = 0 \tag{5.16}$$

is called the characteristic equation of system (5.1) with respect to the solution $\bar{v}$.
Evidently by theorem 5.3 condition (iii) in theorem 5.2 can be replaced by the equivalent condition

(iii') the characteristic equation (5.16) has no complex roots λ with $\mathrm{Re}\,\lambda \geq 0$.

It seems to be an open question whether a steady state $\bar{v}$ of system (5.1) is unstable (i.e. not stable) whenever the characteristic equation has a root with positive real part. However the following considerations show that this problem is of rather academic nature.

Assume that f and g in the linear system (5.8) are differentiable. Then the solutions of this system obey

$$\dot{z}(t)=g(0)z(t)+\int_0^t \dot{g}(t-t')z(t')dt' +\dot{f}(t) . \tag{5.17}$$

It has been shown in (Miller, 1974) that the zero solution of a linear integrodifferential system

$$\dot{z}(t)=Ax(t) + \int_0^t B(t-s)x(s)ds + h(t)$$

is not stable if the equation

$$\det(\lambda I-A-B^*(\lambda)) = 0 \tag{5.18}$$

has a root in the right half of the complex plane. Whether the same condition holds for nonlinear perturbations of linear integrodifferential equations is still unsettled in the general case, but has been shown to be true if certain conditions on the number of roots in the right half plane and their multiplicities are satisfied (Miller, 1974; Cushing, 1975).

Now in the case of equ. (5.17) the relation (5.18) reduces to

$$0=\det(\lambda I-g(0)-(\dot{g})^*(\lambda))=\det(\lambda I-\lambda g^*(\lambda))=\lambda^n \det(I-g^*(\lambda))$$

which coincides with the characteristic equation of the integral system (the factor λ^n plays no role since the associated n-fold root $\lambda=0$ corresponds to the n integration constants appearing in the transition from the integro-differential equation to the original integral equation, in which these constants are not present).

These considerations justify defining the concept of instability in the following way.

Definition 5.4. Let the kernel h be integrable, i.e. $\int_0^\infty \|h(t)\|\,dt<\infty$. Then the solution $\bar{v}$ of system (5.1) is called unstable,

if and only if the characteristic equation $\det(I-g^*(\lambda))=0$ has at least one root with positive real part.
In all special examples throughout this study the notions "unstable" and "not stable" are exchangeable.

5.3 Some simple general criteria for asymptotic stability

According to theorem 5.2, condition (iii') , and definition 5.4 the stability in the neighborhood of a steady state $\bar{v}$ of system (5.1) is essentially determined by the roots of the characteristic equation

$$D(\lambda) = \det(I-g^*(\lambda)) = 0 \quad , \tag{5.19}$$

where

$$g^*(\lambda) = \int_0^\infty e^{-\lambda t} g(t)dt, \; g(t)=(U'_{ij}(\bar{v}_j)h_{ij}(t))_{i,j=1,2,\dots,n}. \tag{5.20}$$

The conditions (i), (ii), and (iv) of theorem 5.2 are assumed to be satisfied in the following. In the remainder of this chapter criteria are developed for when roots of equ.(5.19) lie to the right or all to the left of the imaginary axis.
For definiteness here as everywhere the normalization convention

$$\int_0^\infty |h_{ij}(t)|\, dt = 1, \qquad U'_{ij}(\bar{v}_j) \geq 0 \tag{5.21}$$

is observed. (Note also that throughout this text $h_{ij}(t)$ is supposed to be either nonnegative or nonpositive for all $t \geq 0$).
Condition (5.21) implies

$$|h^*_{ij}(\lambda)| \leq 1 \quad \text{if } \operatorname{Re}\lambda \geq 0 \quad .$$

Hence it is easily concluded that the matrix $I-g^*(\lambda)$ is nonsingular for λ , Re $\lambda \geq 0$ if all eigenvalues of the matrix $(U'_{ij}(\bar{v}_j))$ have a modulus smaller than 1. Thus it is proved:

<u>Theorem 5.5.</u> The steady state $\bar{v}=(\bar{v}_1,\ldots,\bar{v}_n)$ of system (5.1) is asymptotically stable if the spectral radius of the matrix $(U'_{ij}(\bar{v}_j))$ is smaller than one.
For details on the spectral radius see remark 4.7.
This theorem gives a sufficient condition for asymptotic stability. However, even in the case of a single neuron (n=1) this condition is not necessary (see section 5.4). Nevertheless for a certain class of networks, namely those containing only excitatory interactions, it can be shown that a spectral radius greater than one implies instability:

<u>Theorem 5.6.</u> Assume $h_{ij}(t) \geq 0$ for $t \geq 0$, $i,j=1,2,\ldots,n$, and the spectral radius of the matrix $(U'_{ij}(\bar{v}_j))$ to be greater than one. Then the solution $\bar{v}$ of system (5.1) is unstable.

<u>Proof.</u> The proof is based on the following results in the theory of nonnegative matrices essentially due to Perron and Frobenius (for details see Varga, 1962): If $A=(a_{ij})$ is an $n \times n$ matrix with nonnegative entries ($a_{ij} \geq 0$), then A has a nonnegative real eigenvalue equal to its spectral radius $\rho(A)$; moreover $\rho(A)$ does not decrease when any entry of A is increased.
Now for each real $s \geq 0$ the matrix $g^*(s)$ is nonnegative and its entries satisfy

$$g^*_{ij}(s_1) \geq g^*_{ij}(s_2) \qquad \text{if } 0 \leq s_1 \leq s_2 \quad .$$

Therefore $g^*(s)$ has the nonnegative eigenvalue $\rho(g^*(s))$, which is a nonincreasing and continuous function of $s \geq 0$.

Since

$$\rho(g^*(0)) = \rho(U'_{ij}(\bar{v}_j)) > 1, \quad \text{and} \lim_{s \to \infty} \rho(g^*(s)) = 0,$$

it follows that there is a positive s_0 such that $g^*(s_0)$ has the eigenvalue 1, i.e. s_0 is a positive root of $D(\lambda) = 0$, Q.E.D.

Some comments on these two theorems should be made.

1. Instabilities in networks of the type investigated in this study can only occur if the spectral radius ρ of the matrix $(U'_{ij}(\bar{v}_j))$ is greater than one. As is apparent from remark 4.7 (with $x_i = 1$)

$$\rho \leq \max_{1 \leq i \leq n} \sum_{j=1}^{n} U'_{ij}(\bar{v}_j) \quad .$$

   Therefore in small networks (i.e. n small) the interactions have to be very strong in order to make instabilities (and in consequence hysteresis phenomena and oscillations) possible. On the other hand in large networks where many neurons are connected with many others even with weak influences between individual neurons instabilities of equilibria are by no means exceptional.

2. As the proof of theorem 5.6 shows an unstable steady state in a network with only excitatory interactions is always associated with a real positive root of the characteristic equation. Therefore it is unlikely to observe small amplitude oscillations around steady states in such networks. Addition of inhibitory influences seems to facilitate the occurence of oscillatory behavior (e.g. limit cycles). Excitatory networks are predestined for hysteresis effects.

The following two sections deal with cases not covered by the last two theorems, but are restricted to very small networks (n=1,2). These and the last section of this chapter serve also as a background for the studies of oscillations in chapter 6.

5.4 Single neurons.

For a single isolated neuron with steady state $\bar{v}$ the characteristic equation is

$$D(\lambda) = 1 - U'(\bar{v})h^*(\lambda) = 0 \quad .$$

Observing the convention (5.21) it is an easy consequence of the theorems 5.5 and 5.6 that $\bar{v}$ is asymptotically stable if $0 \leq U'(\bar{v}) \leq 1$, and that $\bar{v}$ is unstable if both $U'(\bar{v}) > 1$ and $h(t) \geq 0$ $(t \geq 0)$. It remains to consider the effects of self-inhibition ($h(t) \leq 0$) when $U'(\bar{v}) > 1$. It seems to be difficult to solve this problem satisfactorily and generally. Therefore the investigation is restricted to two types of kernel functions h, where the analysis can be carried out completely and which appear to be important for applications. The first type is given by

$$h_k(t) = -\alpha^k t^{k-1} \exp(-\alpha t)/(k-1)! \quad , \quad t \geq 0 \qquad (5.22)$$

with $k=1,2,\ldots,\alpha > 0$, see fig.5.1

Each of these functions has a single minimum located at $t=(k-1)/\alpha$. With fixed decay rate α the minimum increases with increasing k. This can be interpreted such that with increasing k the main influence of self-inhibition is more and more delayed.

A genuine delay (of length τ) is incorporated in the second

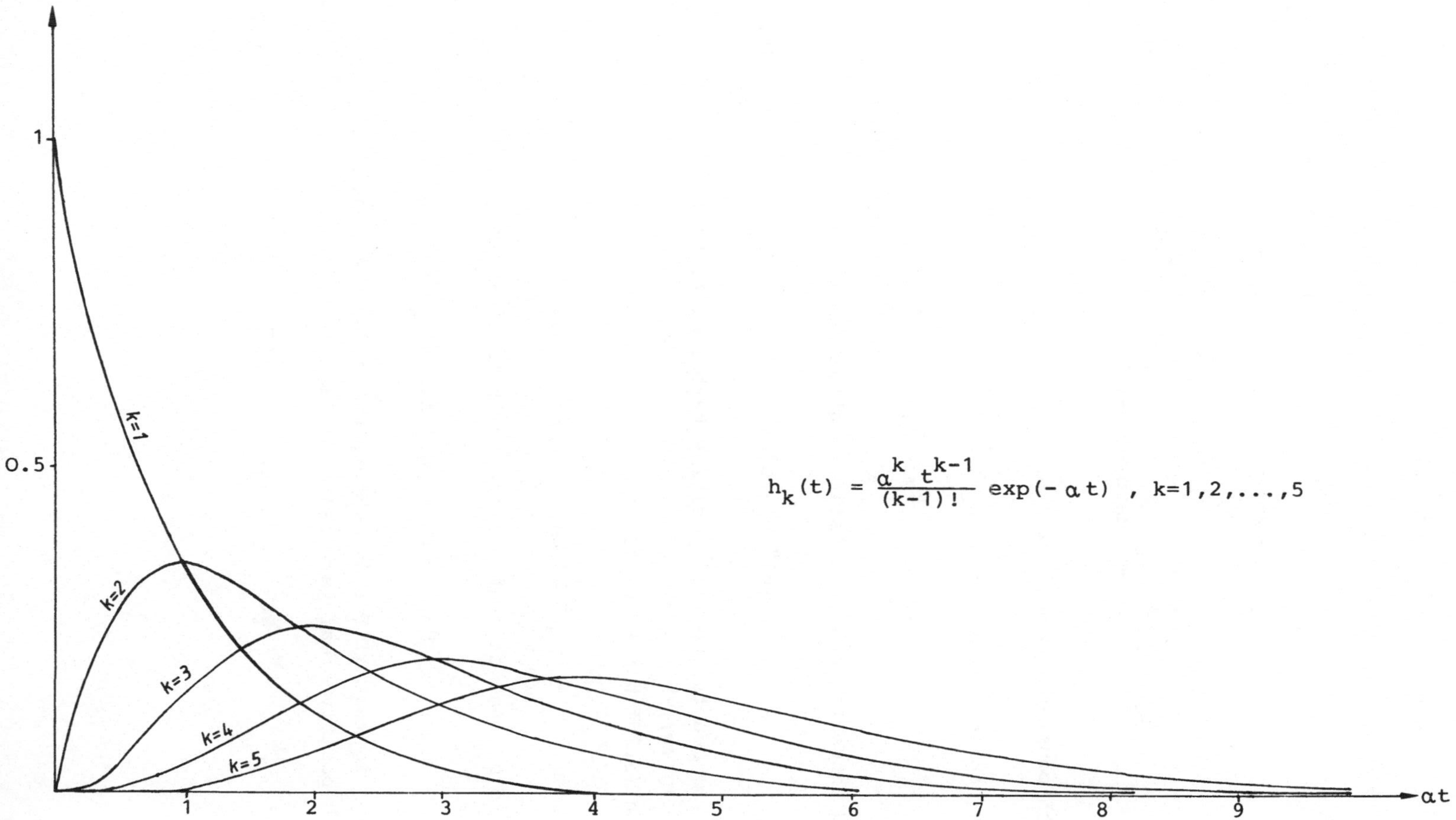

Fig.5.1. Temporal weight functions h_k, k=1,2,...,5

type of weight functions described by

$$h_a(t) = \begin{cases} 0 & , \text{ if } 0 \leq t \leq \tau \\ -\alpha \exp(-\alpha(t-\tau)) & , \text{ if } t \geq \tau \ , \end{cases} \tag{5.23}$$

$$h_b(t) = \begin{cases} 0 & , \text{ if } 0 \leq t \leq \tau \\ -\alpha^2 \ t\exp(-\alpha(t-\tau)) & , \text{ if } t \geq \tau \ , \end{cases} \tag{5.24}$$

with constants $\tau \geq 0$, $\alpha > 0$.

The Laplace transform of h_k is

$$h_k^* = -\alpha^k / (\lambda+\alpha)^k \quad ,$$

and hence the characteristic equation can be written

$$(\lambda+\alpha)^k = - U'(\bar{v})\, \alpha^k \quad . \tag{5.25}$$

Let $\varkappa$ denote the real positive k-th root of $U'(\bar{v})$, $\varkappa = (U'(\bar{v}))^{1/k}$.
Then (5.25) has k solutions

$$\lambda_j = \alpha(\varkappa \rho_j - 1), \quad j=1,2,\ldots,k,$$

where ρ_j runs through all complex k-th roots of -1,

$$\rho_j = \cos(\frac{2j-1}{k}\pi) + i \sin(\frac{2j-1}{k}\pi) \quad .$$

Re ρ_j is maximal for j=1. Therefore a root of equ.(5.25) with positive real part exists if and only if

$$k>2 \text{ and } U'(\bar{v}) > (\cos\frac{\pi}{k})^{-k} \quad . \tag{5.26}$$

In summary:

Theorem 5.7. The steady state $\bar{V}$ of a single isolated neuron with a self-inhibitory weight function h_k given by equ. (5.22) is unstable if the inequalities (5.26) hold. If one of these inequalities is reversed then the steady state is asymptotically stable.

The table below lists some values of

$$p(k) = [\cos(\pi/k)]^{-k} \quad :$$

k	2	3	4	6	8	10	16	24	32
p(k)	∞	8	4	2.37	1.88	1.65	1.36	1.23	1.17

For k=1 and k=2 the steady state can never be unstable even with arbitrary large $U'(\bar{V})$. Note however, that for k=2 the roots $\lambda_{1,2}$ have non-vanishing imaginary parts (if $\varkappa \neq 0$). Therefore damped oscillations can be expected in a neighborhood of $\bar{V}$.

It is easily seen from the table (more precisely from the fact that p(k) decreases with increasing k) and the inequalities (5.26) that the equilibrium is destabilized with a smaller slope of U the more the main weight of the inhibitory influence is shifted to the past.

This observation will be strengthened by the investigation of the delay kernels h_a and h_b in (5.23), (5.24).

All considerations on stability in section 5.2, in particular theorem 5.2, remain true if the condition (i) of theorem 5.2 is weakened to also admit functions h_{ij} with delays $\tau_{ij} \geq 0$, i.e.

$$h_{ij} = \begin{cases} 0 & 0 \leq t < \tau_{ij} \\ \text{continuous for } t \geq \tau_{ij} \, . \end{cases}$$

The Laplace transforms of h_a and h_b are

$$h_a^* = -\alpha\ e^{-\lambda t}\ /(\lambda+\alpha)\quad ,$$

$$h_b^* = -\alpha^2 e^{-\lambda t}\ /(\lambda+\alpha)^2\ .$$

Therefore the characteristic equations can be written as

$$D_a(\lambda) = \lambda + \alpha + \alpha U'(\bar{v})e^{-\lambda\tau} = 0 \tag{5.27}$$

and

$$D_b(\lambda) = (\lambda+\alpha)^2 + \alpha^2 U'(\bar{v})e^{-\lambda\tau} = 0 \tag{5.28}$$

respectively.

(5.27) and (5.28) are transcendental equations having infinitely many roots in the complex plane. Criteria when there are roots with positive real parts or when all roots have negative real parts are well known:

<u>Theorem 5.8.</u> (Bellman-Cooke, 1963, chap.13).

a) All complex roots z of the equation $(p-z)e^z+q=0$, where p and q are real, have negative real parts if and only if

   (α) $p<1$, and

   (β) $p<-q<\sqrt{a_1^2+p^2}$,

   where a_1 is the root of $a=p\tan a$ such that $0<a<\pi$ (if $p=0$, then $a_1=\pi/2$).

b) Let $H(z)=(z^2+pz+q)e^z+r$, where p is real and positive, q is real and nonnegative, and r is real. Let $a_o=0$ and a_k ($k\geq 1$) the unique root of the equation

$$\operatorname{ctg} a = (a^2-q)/(ap)$$

   which lies in the interval $(k\pi-\pi, k\pi)$. A number w is defined as follows:

(α) if $r \geq 0$ and $p^2 \geq 2q$, $w = 1$,

(β) if $r \geq 0$ and $p^2 < 2q$, w is the odd k for which a_k lies closest to $\sqrt{q-p^2/2}$,

(γ) if $r < 0$ and $p^2 \geq 2q$, $w = 2$,

(δ) if $r < 0$ and $p^2 < 2q$, w is the even k for which a_k lies closest to $\sqrt{q-p^2/2}$.

A necessary and sufficient condition that all the roots of $H(z)=0$ lie to the left of the imaginary axis is that

(i) $r \geq 0$ and $(r \sin a_w)/(pa_w) < 1$, or

(ii) $-q < r < 0$ and $(r \sin a_w)/(pa_w) < 1$.

(Remark: there are minor errors in the formulation of part b in the reference). This theorem can be proved by using the method of Čebotarev and Pontrjagin (see Bellman-Cooke, 1963 , chap.13).

Applying theorem 5.8 to the characteristic equations $D_a(\lambda)=0$ and $D_b(\lambda)=0$ the following result is obtained.

Corollary 5.9.a) The steady state $\bar{v}$ of a single isolated self-inhibitory neuron with a temporal weight function h_a given by (5.23) is asymptotically stable if

$$[(U'(\bar{v}))^2-1] \ (\alpha\tau)^2 < a_1^2 \quad , \tag{5.29}$$

where a_1 denotes the solution of $a = -\alpha\tau \tan a$ in the interval $\pi/2 < a < \pi$. If the inequality (5.29) is reversed $\bar{v}$ is unstable.

b) In case of the weight function h_b (5.24) the steady state $\bar{v}$ is asymptotically stable if and only if

$$[U'(\bar{v}) -1] \ (\alpha\tau)^2 < \tilde{a}_1^2 \quad , \tag{5.30}$$

where $\tilde{a}_1$ solves $\cot a = (\frac{a}{\alpha\tau} - \frac{\alpha\tau}{a})/2$ in the interval $0 < a < \pi$.

An obvious conclusion from this corollary is that given arbitrary $\alpha > 0$, $U'(\bar{v}) > 1$ then the steady state is unstable provided the delay is sufficiently large.

There seems to be no simple general criterion for the instability of the steady state in case of self-inhibition. However it is possible to derive a general bifurcation theorem which may sometimes be used to find local regions in parameter space associated with unstable equilibria. The parameter is here $\varkappa = U'(\bar{v})$. In order to formulate the theorem simply, for each $\omega \in \mathbb{R}$ three quantities are defined:

$$R(\omega) = \int_0^\infty h(t) \cos \omega t \, dt, \qquad I(\omega) = \int_0^\infty h(t) \sin \omega t \, dt ,$$

and $$G(\omega) = \int_0^\infty t h(t) \cos \omega t \, dt \quad .$$

<u>Theorem 5.10.</u> Suppose $\int_0^\infty |h(t)| \, dt < \infty$ and $\int_0^\infty |t h(t)| \, dt < \infty$. If there is $\omega_0 > 0$ such that

$$R(\omega_0) > 0, \; I(\omega_0) = 0 \quad ,$$

then the steady state $\bar{v}$ of equ. (5.1), n=1, is unstable if

$$U'(\bar{v}) \in (1/R(\omega_0) - \varepsilon \, , 1/R(\omega_0)), \qquad G(\omega_0) < 0$$

or

$$U'(\bar{v}) \in (1/R(\omega_0), \; 1/R(\omega_0) + \varepsilon), \qquad G(\omega_0) > 0$$

where ε is a sufficiently small positive number.

<u>Proof.</u> With the parameter $\varkappa = U'(\bar{v})$ the characteristic equation (5.19) can be written

$$D(\lambda, \varkappa) = 1 - \varkappa h^*(\lambda) = 0 \quad . \tag{5.31}$$

(5.31) has a purely imaginary root $i\omega_o$, $\omega_o > 0$ for $\varkappa = \varkappa_o$ if and only if

$$D(i\omega_o, \varkappa_o) = 1-\varkappa_o(R(\omega_o)-iI(\omega_o)) = 0 \quad .$$

This equation can be satisfied for some $\varkappa_o > 0$ if and only if

$$I(\omega_o) = 0 \text{ and } R(\omega_o) > 0$$

for some $\omega_o > 0$. In this case $\varkappa_o = 1/R(\omega_o)$.
If the derivative $\partial D(\lambda,\varkappa)/\partial\lambda$ does not vanish at $(i\omega_o, \varkappa_o)$, i.e. if

$$\varkappa_o(G(\omega_o) - i\int_0^\infty th(t)\sin\omega_o t dt) \neq 0,$$

then the implicit function theorem can be applied to the equation $D(\lambda,\varkappa)=0$ to ensure the existence of a function $\lambda(\varkappa)$, defined in a neighborhood of $\varkappa_o$ and obeying

$$D(\lambda(\varkappa),\varkappa) = 0, \quad \lambda(\varkappa_o) = i\omega_o \quad .$$

It follows from

$$0 = \frac{dD(\lambda(\varkappa),\varkappa)}{d\varkappa} = \frac{\partial D(\lambda(\varkappa),\varkappa)}{\partial\lambda}\frac{d\lambda(\varkappa)}{d\varkappa} + \frac{\partial D(\lambda(\varkappa),\varkappa)}{\partial\varkappa}$$

and (5.31) that

$$\text{Re }\lambda'(\varkappa_o) = G(\omega_o)/|\partial D(i\omega_o,\varkappa_o)/\partial\lambda|^2 \quad .$$

The sign of this last quantity determines the direction of bifurcation of unstable solutions from purely imaginary roots as described in the theorem. Q.E.D.

In applications the kernel function $h(t)$ often can be approximated by sums of exponentials,

$$h(t) = \sum_{k=1}^{m} \alpha_k \exp(-\gamma_k t), \quad \gamma_k > 0 \quad .$$

The investigation of the condition $I(\omega) = 0$ in the theorem then reduces to the evaluation of the real zeros of a polynomial since in this case

$$I(\omega) = \sum_{k=1}^{m} \alpha_k \frac{\omega}{\gamma_k^2 + \omega^2} ,$$

$$R(\omega) = \sum_{k=1}^{m} \alpha_k \frac{\gamma_k}{\gamma_k^2 + \omega^2} ,$$

$$G(\omega) = \sum_{k=1}^{m} \alpha_k \frac{\gamma_k^2 - \omega^2}{(\gamma_k^2 + \omega^2)^2} .$$

5.5 Pairs of neurons

The characteristic equation, determining the stability behavior in the neighborhood of an equilibrium $\bar{v}=(\bar{v}_1,\bar{v}_2)$ for a pair of neurons, is

$$0=D(\lambda)=1-k^*_{11}(\lambda)-k^*_{22}(\lambda)+k^*_{11}(\lambda)k^*_{22}(\lambda)-k^*_{12}(\lambda)k^*_{21}(\lambda), \tag{5.32}$$

where

$$k^*_{ij}(\lambda)=U'_{ij}(\bar{v}) \int_0^{\infty} e^{-\lambda t} h_{ij}(t)dt .$$

It is difficult to obtain a complete survey on the location of the roots of $D(\lambda)$ for arbitrary weight functions h_{ij}. However, two conclusions can be drawn from the general results in sect. 5.3:

(i) <u>if the largest eigenvalue of the matrix</u>

$$\begin{pmatrix} U'_{11}(\bar{v}_1) & U'_{12}(\bar{v}_2) \\ U'_{21}(\bar{v}_1) & U'_{22}(\bar{v}_2) \end{pmatrix} \tag{5.33}$$

<u>is smaller than one the steady state is asymptotically stable,</u>

(ii) <u>if all interconnections are excitatory (i.e. $H_{ij} \in \{0,1\}$ for all i,j) and the largest eigenvalue of (5.33) is greater than one then the steady state is unstable</u>.

There is another case which can be solved simply and satisfactorily. Assume the two neurons are mutually inhibitory ($H_{12}=H_{21}= -1$) and there are no self-interactions ($U_{ii}(\bar{v}_i)=0$). Then

(iii) <u>the steady state $\bar{v}$ is unstable if $U'_{12}(\bar{v}_2)\cdot U'_{21}(\bar{v}_1) > 1$.</u>

(Note that in this case the last product is just the spectral radius of the matrix (5.33)).

<u>Proof of (iii).</u> For real nonnegative s the characteristic equation is

$$D(s)=1-U'_{12}(\bar{v}_2)\; U'_{21}(\bar{v}_1)F(s) = 0$$

with

$$F(s)=\int_0^\infty e^{-st}|h_{12}(t)|\,dt\cdot\int_0^\infty e^{-st}|h_{21}(t)|\,dt \quad .$$

F(s) is a continuous function of s satisfying

$$F(0)=1 \quad \text{and} \quad \lim_{s\to\infty} F(s)=0 .$$

Therefore there is a value $s=s_0>0$ with $D(s_0)=0$. Q.E.D.

More difficulties arise in the case of an inhibitory - excitatory coupled pair of neurons, $H_{12}=1$, $H_{21}= -1$. Of course the equilibrium $\bar{v}=(\bar{v}_1,\bar{v}_2)$ is asymptotically stable if $U'_{12}(\bar{v}_2)\cdot U'_{21}(\bar{v}_1)< 1$. However, for the reversed inequality the stability depends critically on the shape of the weight functions h_{12} and h_{21}.

Consider first kernels of the following type:

$$h_{12}(t)=T_1 t^{k_1-1} e^{-\alpha_1 t} \tag{5.34 a}$$

$$h_{21}(t)=T_2 t^{k_2-1} e^{-\alpha_2 t} \quad , \quad t \geq 0, \tag{5.34 b}$$

with constants $k_i \in \mathbb{N}, \alpha_i > 0$ and T_i such that $\int_0^\infty |h_{ij}(t)|\,dt=1$,

$$T_1 = \alpha_1^{k_1} / (k_1-1)! \; , \quad T_2 = -\alpha_2^{k_2}/(k_2-1)! \; .$$

Then the characteristic equation takes the form

$$1+U'_{12}(\bar{v}_2)U'_{21}(\bar{v}_1)\,\alpha_1^{k_1} \cdot \alpha_2^{k_2} / [(\lambda+\alpha_1)^{k_1} (\lambda+\alpha_2)^{k_2}]=0 \; . \tag{5.35}$$

Defining

$$G(\lambda) = -\alpha_1^{k_1} \alpha_2^{k_2} / [(\lambda+\alpha_1)^{k_1} (\lambda+\alpha_2)^{k_2}] \quad ,$$

and

$$U' = U'_{12}(\bar{v}_2) \cdot U'_{21}(\bar{v}_1) \quad ,$$

the condition that all roots of equ. (5.35) lie to the left of the imaginary axis can be formulated as

$$1-U'G(\lambda) \neq 0 \quad \text{for Re}\,\lambda \geq 0 \quad . \tag{5.36}$$

By the argument principle, (5.36) is satisfied if and only if

$$\omega \geq 0 \text{ and Im } G(i\omega)=0 \text{ implies } 1-U' \text{ Re } G(i\omega) > 0. \tag{5.37}$$

With $M^{-1} = \max\{\text{Re } G(i\omega) : \omega \geq 0, \text{ Im } G(i\omega) = 0\}$ condition (5.37) can be expressed as

$$M > U' \quad .$$

In order to determine M observe that

$$G(i\omega) = - \frac{\alpha_1^{k_1} \alpha_2^{k_2}}{(\alpha_1^2 + \omega^2)^{k_1/2} (\alpha_2^2 + \omega^2)^{k_2/2}} \cdot e^{-i(k_1\Theta_1(\omega) + k_2\Theta_2(\omega))},$$

where $\Theta_i(\omega) = \tan^{-1}(\omega/\alpha_i)$.

Since $G(0) = -1$ and $|G(i\omega)|$ decreases with increasing ω the maximum of Re $G(i\omega)$ under the condition Im $G(i\omega)=0$ occurs at the first positive $\omega=\omega_0$ satisfying Im $G(i\omega_0)=0$. Obviously ω_0 is the solution of

$$k_1\Theta_1(\omega_0) + k_2\Theta_2(\omega_0) = \pi .$$

Then

(iv) <u>according as</u>

$$M = 1/\text{Re } G(i\omega_0) = (\alpha_1^2 + \omega_0^2)^{k_1/2} (\alpha_2^2 + \omega_0^2)^{k_2/2} / (\alpha_1^{k_1} \alpha_2^{k_2})$$

<u>is larger or smaller than U' the steady state is asymptotically stable or unstable respectively.</u>

For a fixed value $\alpha_1^{k_1} \alpha_2^{k_2}$ and fixed $k=k_1+k_2$ it follows from symmetry considerations that $M=M(\alpha_1, \alpha_2, k_1, k_2)$ is minimal for $\alpha = \alpha_1 = \alpha_2$, $k_1 = k_2$.

Then $\Theta_1(\omega_0) = \Theta_2(\omega_0) = \pi/k$, and

$$M = M_{min} = \left(\frac{\sqrt{\alpha^2 + \omega_0^2}}{\alpha}\right)^k .$$

Since $\tan \Theta_i(\omega_0) = \omega_0/\alpha$,

$$M_{min} = (\cos \Theta_1(\omega_0))^{-k} = (\cos(\pi/k))^{-k} ,$$

a quantity which is independent of α.

In summary the following result has been obtained.

Theorem 5.11. Let be given a pair of excitatory-inhibitory coupled neurons with weight functions of the form (5.34 a), (5.34 b) and $h_{ii}(t)=0$ $(i=1,2)$. Let $k=k_1+k_2$. Then the steady state $\bar{v}=(\bar{v}_1,\bar{v}_2)$ is asymptotically stable if

$$U'_{12}(\bar{v}_2)\cdot U'_{21}(\bar{v}_1) < [\cos(\pi/k)]^{-k} \quad .$$

However if the reverse inequality holds then $\bar{v}$ is unstable provided $|\alpha_1-\alpha_2|<\varepsilon$ for some sufficiently small $\varepsilon>0$.

For specific values of $[\cos\pi/k]^{-k}$ see the table below theorem 5.7.

In many theoretical investigations the kernels (5.34 a) and (5.34 b) are assumed to be of purely exponential type, i.e. $k_1=k_2=1$. Then,as the previous theorem and the table show, the steady state cannot be unstable. On the other hand the same theorem and table demonstrate that a pair of excitatory - inhibitory coupled neurons is not always a stable configuration.Comparing the results for single and pairs of neurons(theorems (5.7),(5.11)) it can be seen that, taking k, k_1, k_2 as measures of response delays in the single, the first, and the second neuron of the pair respectively, for instability in the single neuron the delay has to equal the sum of the delays in the pair.
This heuristic principle will be generalized in sect. 5.6.
The second class of kernel functions to be considered here is defined by

$$i \neq j : \qquad h_{ij}(t) = \begin{cases} 0 & \text{for } 0\leq t<\tau_{ij} \\ H_{ij}\alpha_i \exp(-\alpha_i(t-\tau_{ij})) & \text{for } t\geq\tau_{ij}, \end{cases} \tag{5.38}$$

$$h_{ii}(t)\equiv 0,$$

with constants $\tau_{ij} \geq 0$, $\alpha_i > 0$, $H_{12}=1$, $H_{21}=-1$.

The characteristic function now is

$$\lambda^2+\lambda(\alpha_1+\alpha_2)+\alpha_1\alpha_2+\alpha_1\alpha_2 \; U'_{12}(\bar{v}_2)U'_{21}(\bar{v}_1)e^{-\lambda(\tau_{12}+\tau_{21})} .$$

Again the stability problem can be solved by help of theorem 5.8 :

<u>Corollary 5.11.</u> Let be given a pair of excitatory-inhibitory coupled neurons with weight functions (5.38) and $h_{ii}\equiv 0$. Let $\tau=\tau_{12}+\tau_{21}$, $U'=U'_{12}(\bar{v}_2)\cdot U'_{21}(\bar{v}_1)$, and $a_1 \in (0,\pi)$ the number satisfying

$$\operatorname{ctg} a_1 = (a_1^2-\tau^2\alpha_1\alpha_2)/(a_1\tau(\alpha_1+\alpha_2)) .$$

Then the steady state $\bar{v}=(\bar{v}_1,\bar{v}_2)$ of the pair is asymptotically stable / unstable if $(\tau^2\alpha_1\alpha_2)^2((U')^2-1) \lesseqgtr a_1^2(a_1^2+\tau^2(\alpha_1^2+\alpha_2^2))$ respectively.

As τ occurs of order 4 on the left hand side and of order 2 on the right hand side of this inequality, the steady state is unstable if $U'>1$ and τ is sufficiently large.

Finally the effects of self-inhibition or -excitation in a pair of neurons will be discussed for some simple examples of kernel functions.
A complete picture is easily obtained if all weighting functions can be represented as

$$h_{ij}=H_{ij}\alpha \exp(-\alpha t), \alpha>0, \quad H_{ij}\in\{-1,0,1\} \quad , i,j=1,2.$$

With the abbreviations

$$a_{ij}=H_{ij}\,U'_{ij}\,(\bar{v}_j)$$

the characteristic equation (5.32) reads

$$\lambda^2+\lambda\alpha(2-a_{11}-a_{22})+\alpha^2(1-a_{11}-a_{22}+a_{11}a_{22}-a_{12}a_{21}) = 0,$$

which has the solutions

$$\alpha^{-1}\lambda_{1,2} = -1+\frac{a_{11}+a_{22}}{2} \pm \frac{1}{2}\sqrt{(a_{11}-a_{22})^2+4a_{12}a_{21}} \quad .$$

There are exactly two cases for which an eigenvalue with positive real part exists:
Either

$$a_{11}+a_{22} > 2 \quad \text{and} \quad (a_{11}-a_{22})^2 + 4a_{12}\, a_{21} < 0, \tag{5.39}$$

or

$$(a_{11}-a_{22})^2 + 4a_{12}a_{21} > 0 \quad \text{and} \quad a_{22}+a_{11}+a_{12}a_{21} > 1+a_{11}a_{22} \quad . \tag{5.40}$$

In the case (5.39) the steady state is an unstable focus, in the case (5.40) a saddle or an unstable node.
A detailed discussion of these conditions in the context of spatially homogeneous networks was given in (Wilson and Cowan, 1973) and (Cowan and Ermentrout, 1978).

As it can be seen from the second inequality in (5.39) a mutually inhibitory pair with purely exponential kernels cannot exhibit undamped oscillations in the neighborhood of the steady state. The simplest configuration allowing this behavior needs a somewhat delayed self-inhibitory response:
Let

$$h_{11}(t)=h_{22}(t)=H_{11}\,\alpha^2 t\, \exp(-\alpha t), \tag{5.41a}$$

$$h_{12}(t)=h_{21}(t)=H_{12}\;\alpha\; \exp(-\alpha t),\ \alpha > 0. \tag{5.41b}$$

If

$$a_{11}=a_{22} \quad \text{and} \quad a_{12}=a_{21} \tag{5.42}$$

the characteristic equation becomes

$$0=((\lambda+\alpha)^2-a_{11}\alpha^2)^2-\alpha^2 a_{12}^2(\lambda+\alpha)^2 ,$$

having solutions

$$\lambda/\alpha = \pm \frac{a_{12}}{2} - 1 \pm \sqrt{a_{12}^2 + 4a_{11}} \tag{5.43}$$

with all 4 possible combinations of signs.

An eigenvalue with positive real part and non-vanishing imaginary part occurs if and only if

$$4 < a_{12}^2 < -4a_{11} \tag{5.44}$$

i.e. there has to be sufficiently strong self-inhibition and either strong lateral excitation or inhibition.
In this situation sustained oscillations can be expected.
A similar condition was derived by Stein et.al., 1974 b.
Originally the idea that reciprocal inhibition can generate oscillations was promoted by Harmon, 1961, and Reiss, 1962, and later by Szekely, 1965.

5.6 Closed chains of neurons

A closed chain of length n consists of n neurons, $n \geq 1$, labeled from 1 to n, such that the i-th neuron exerts an inhibitory or excitatory influence on the i+1-th neuron, where $i=1,2,\ldots,n$, the index $n+1$ being identified with the index 1 throughout this section.
Other connections between the cells are excluded, also any self-inhibitory or -excitatory effects. Assume this network receives a temporally constant input $e=(e_1,\ldots,e_n)$ and $\bar{v}=(\bar{v}_1,\ldots,\bar{v}_n)=\bar{v}(e)$ is a steady state of the net belonging to e.
Further it is assumed that the temporal weighting functions

have the form

$$h_{i+1,i}=H_{i+1,i}\alpha_i^{k_i}\cdot t^{k_i-1}\cdot \exp(-\alpha_i t)/(k_i-1)!,\ i=1,2,\dots,n \tag{5.45}$$

with $\quad H_{i+1,i}\in\{-1,1\}\quad,\quad t\geq 0,\ k_i\in\mathbb{N}\quad,\alpha_i>0\quad.$

As the maximum of $h_{i+1,i}$ is located at $t=(k_i-1)/\alpha_i$ the exponents k_i can be used to model various delays in the response of the following neuron to the preceding one. With these assumptions, since

$$h^*_{i+1,i}=H_{i+1,i}\,\alpha_i^{k_i}/(\lambda+\alpha_i)^{k_i}\quad,$$

the characteristic equation becomes

$$0=D(\lambda)=1-\prod_{i=1}^{n}H_{i+1,i}\,U'_{i+1,i}(\bar{v}_i)\alpha_i^{k_i}/(\lambda+\alpha_i)^{k_i}\quad. \tag{5.46}$$

With the abbreviations

$$U'=\prod_{i=1}^{n}H_{i+1,i}\,U'_{i+1,i}(\bar{v}_i)\quad,$$

$$G(\lambda)=\prod_{i=1}^{n}\alpha_i^{k_i}/(\lambda+\alpha_i)^{k_i}\quad,$$

equ.(5.46) reads

$$1-U'\,G(\lambda)=0\quad. \tag{5.47}$$

Suppose first that $U'\geq 0$. Then it follows from

$$|G(\lambda)|\leq 1\qquad\text{if }\mathrm{Re}\,\lambda\geq 0$$

that equ.(5.46) cannot have a root with positive real part if $U'<1$. On the other hand if $U'>1$ then there is a real positive root of (5.47) as consequence of

$$\lim_{s \to \infty} G(s)=0, \quad s \in \mathbb{R},$$

and

$$G(0) = 1 \quad .$$

Now let $U' < 0$. A slight generalization of the arguments leading to theorem 5.11 will imply the second part of the following statement.

Theorem 5.12. Let be given a closed chain of neurons with weighting functions of the form (5.45). Then

(i) if $U' > 1$ the steady state $\bar{v}$ is unstable,

(ii) if $-p(k) < U' < 1$ then $\bar{v}$ is asymptotically stable,

where $k = \sum_{i=1}^{n} k_i$, $p(k) = [\cos(\pi/k)]^{-k}$ $(k > 2)$, $p(1)=p(2)=\infty$,

(iii) if $U' < -p(k)$ then there is a number $\varepsilon > 0$ such that $\bar{v}$ is unstable whenever

$$|\alpha_i - \alpha_j| < \varepsilon \quad , \quad i,j=1,2,\ldots,n \quad .$$

Remark 1. Theorem 5.7 represents a special case of this theorem where the chain length is n=1; theorem 5.11 represents the special case n=2 and $U' < 0$.

Remark 2. For values of p(k) see the table below theorem 5.7.

Remark 3. The condition on α_i in part (iii) of the theorem is satisfied if all decay rates are equal. For this situation the theorem gives a complete solution of the local stability problem.

Remark 4. There are qualitatively different types of instability in condition (i) and (iii) of the theorem. In case of condition

(i) a real positive eigenvalue exists, moreover it can be shown that if the functions U_{ij} are bounded then at least two further steady states exist. Therefore it can be conjectured that most solutions starting near $\bar{v}$ will tend to one of the other steady states (hysteresis). In case of condition (iii) an eigenvalue with positive real and non-vanishing imaginary part exists and the steady state is unique (as is easily concluded from theorem 4.13 since $\prod_{i=1}^{n} H_{i+1,i} = -1$). Therefore most solutions near the steady state $\bar{v}$ will perform oscillations with increasing amplitude.

<u>Remark 5.</u> Let all couplings in the chain be inhibitory, i.e. $H_{i+1,i} = -1$. Then the sign of U' is positive if n is even, and it is negative if n is odd. According to the previous remark the chains with odd n can serve as neural oscillators. Those with even n (assuming $U' > 1$) exhibit hysteresis phenomena, which in view of example 1.2 can be very complex.

6. Oscillations in neural networks

6.1 Introduction

Oscillations represent a very common feature on all levels of nervous activity. Starting from the repetitive firing of spontaneously active single cells (Eccles,1973), the next complex phenomenon consists in regularly distributed bursts of impulses in isolated cells (Chen et.al.1971; Gainer,1972). In this case the impulse frequency changes periodically. Oscillations due to the interaction of a few nerve cells seem to be established experimentally e.g. in the stomatogastric ganglion of the spiny lobster (Warshaw & Hartline, 1976) and in the locomotory system of the leech (Friesen & Stent, 1977). At a higher level oscillations of whole cell populations are known, e.g. in the olfactory bulb (Freeman,1975) or those reflected by the EEG. Two further important examples are the respiratory and circulatory systems, the rhythms of which are governed by neural activity.

It is impossible to give a complete survey either of the experimental or of the theoretical literature. More urgent is a basic and systematic approach to this field.

The emphasis in this chapter is on mathematical principles which give insight into how oscillations, in particular periodic oscillations, can be generated by the interaction of elements which, when isolated, exhibit no periodic changes.

The main difference from earlier treatments of these problems consists in the use of higher dimensional analytical techniques in nonlinear systems.

6.2 Oscillations in closed chains of neurons

According to the definition of a closed chain of length $n \geq 1$ in sect. 5.6 the dynamic equations have the form

$$v_i = v_{io} + U_{i,i-1}(v_{i-1}) * h_{i,i-1} + \bar{S}_i(E_i) * \bar{h}_i, \quad i=1,2,\ldots,n. \qquad (6.1)$$

(compare equ. (5.2)). Here and in the following the index O is identified with the index n.

It is assumed that the input $E(t)=(E_1(t),\dots,E_n(t))$ to the system is temporally constant, $E(t)=\bar{E}$.
Let $\bar{v}=(\bar{v}_1,\dots,\bar{v}_n)=\bar{v}(\bar{E})$ be a steady state solution of system (6.1) corresponding to the input $\bar{E}$. Then the perturbed variables

$$w_i(t)=v_i(t)-\bar{v}_i \tag{6.2}$$

satisfy (see equ.(5.2))

$$w_i=H_i\cdot U_i(w_{i-1})*h_i,\quad i=1,2,\dots,n, \tag{6.3}$$

where the simplifying notations

$$H_i=H_{i,i-1}=\int_0^\infty h_{i,i-1}(t)dt\in\{-1,1\},$$

$$U_i(w_{i-1})=U_{i,i-1}(w_{i-1}+\bar{v}_{i-1})-U_{i,i-1}(\bar{v}_{i-1}),$$

and

$$h_i(t) = |h_{i,i-1}(t)|$$

have been used. Note that system (6.3) is equivalent to system (6.1). If the functions $U_{i,i-1}$ are strictly increasing then obviously

$$\xi U_i(\xi)>0 \quad \text{for all real } \xi \neq 0. \tag{6.4}$$

Without loss of generality it can always be assumed that $H_1\in\{-1,1\}$ and $H_2=H_3=\dots=H_n=1$, as is shown in the following

Lemma 6.1. Assume the condition (6.4) is satisfied for $i=1,2,\dots,n$. Then system (6.3) is equivalent to a system

$$\tilde{w}_1=(\prod_{k=1}^{n} H_k)U_1(\tilde{w}_n)*h_1,$$

$$\tilde{w}_i=\bar{U}_i(\tilde{w}_{i-1})*h_i, \quad i=2,\dots,n,$$

satisfying

$$\xi\,\bar{U}_i(\xi)>0 \text{ for all real } \xi\neq 0,\ i=2,\dots,n.$$

The relation between the variables w_i and $\tilde{w}_i$ is given by formula (6.5).

<u>Proof.</u> Substitution of the transformed quantities

$$\tilde{w}_i=\Big(\prod_{k=i+1}^{n} H_k\Big)w_i, \qquad \Big(\prod_{n+1}^{n} H_k=1\Big) \tag{6.5}$$

$$\bar{U}_i\,(\tilde{w}_{i-1})=\Big(\prod_{k=i}^{n} H_k\Big)U_i(w_{i-1})$$

into the equs.(6.3) results in

$$\tilde{w}_i=\Big(\prod_{k=i}^{n}H_k\Big)\ U_i(w_{i-1})*h_i=\bar{U}_i(\tilde{w}_{i-1})*h_i, \ i=1,\dots,n.$$

Because of (6.4)

$$\xi\,\bar{U}_i(\xi)=\xi\Big(\prod_{k=i}^{n} H_k\Big)U_i\Big(\Big(\prod_{k=i}^{n} H_k\Big)\xi\Big)>0, \ i=2,3,\dots,n,$$

$$\bar{U}_1(\xi)=\Big(\prod_{k=1}^{n} H_k\Big)U_1(\xi), \text{ Q.E.D.}$$

The preceeding lemma justifies calling (6.1) a <u>repressible system</u> if the product $\prod_{k=1}^{n} H_k$ of coupling signs is negative, and calling it an <u>inducible system</u> if this product is positive. This terminology is in line with that used for two models of cellular control processes which have been the object of intense mathematical investigations during recent years (Goodwin,1965; Griffith,1968; Tyson,1975; Hastings-Tyson-Webster,1977; Tyson-Othmer,1978).
These models describe the dynamic behavior in closed loops of enzyme-catalyzed reaction sequences within living cells.

Fortunately a theorem of Hastings,Tyson, and Webster (1977), is applicable to a certain class ofclosed chains of neurons as follows.
Just as in sect. 5.6 assume the weight functions to have the form

$$h_i(t)=|h_{i,i-1}(t)| =\alpha_i^{k_i} t^{k_i-1} \exp(-\alpha_i t)/(k_i-1)!, \tag{6.6}$$

$\alpha_i > 0$, $k_i \in \mathbb{N}$.

The choice of such kernel functions allows one to transform the system (6.3) of integral equations into a system of differential equations.
Define

$$h_i^{(m)}(t)=\alpha_i^m \, t^{m-1} \exp(-\alpha_i t)/(m-1)!,\ m=1,2,\ldots,k_i, \tag{6.7}$$

$$h_i^{(0)}(t) =\alpha_i^{-1} \delta(t), \tag{6.8}$$

$$x_{im}=H_i U_i(w_{i-1}) * h_i^{(m)} \quad , \tag{6.9}$$

(in particular $x_{io}(t)=H_i U_i(w_{i-1}(t))/\alpha_i$).

Note that

$$w_i = x_{ik_i} \quad . \tag{6.10}$$

It follows by differentiating both sides of equ. (6.9) that the functions x_{im} obey the differential equations

$$\dot{x}_{im} = dx_{im}/dt=\alpha_i \, x_{i,m-1}(t)-\alpha_i x_{im}(t), \tag{6.11}$$

$i=1,\ldots,n;\ m=1,\ldots,k_i$.
According to lemma 6.1 it is assumed that $H_2=H_3=\ldots=H_n=1$.

Then defining functions $y_1,\ldots,y_p$, $(p=\sum_{i=1}^{n} k_i)$, by

$$y_j = x_{im}, \text{ if } j=m+\sum_{l<i} k_l \text{ , } j=1,2,\ldots,p, \tag{6.12}$$

(in particular $w_i=y_j$ with $j=k_1+\ldots+k_i$),

the final form of the differential equation system is obtained:

$$\dot{y}_1=H_1U_1(y_p)-\alpha_1y_1, \tag{6.13a}$$

$$\dot{y}_j=\alpha_iy_{j-1}-\alpha_iy_j, \text{ if } j=m+\sum_{l<i} k_l, \ 1<m\leq k_i, \tag{6.13b}$$

$$\dot{y}_j=U_i(y_{j-1})-\alpha_iy_j, \text{ if } j=1+\sum_{l<i} k_l, \ i>1, \tag{6.13c}$$

$j=2,3,\ldots,p.$

The essential property of this system is that the time derivative of y_j depends only on y_j itself and on the preceding variable y_{j-1} (with the identification $y_0=y_p$). Therefore (6.13) turns out to be a special case of the following system investigated by Hastings, Tyson, Webster, 1977:

$$\dot{y}_1=g_1(y_p,y_1) \tag{6.14a}$$

$$\dot{y}_j=g_j(y_{j-1},y_j), \quad 2\leq j\leq p. \tag{6.14b}$$

They assume the following conditions to hold for the functions g_j:

$$g_j(0,0) = 0, \qquad j=1,\ldots,p. \tag{6.15}$$

Moreover there are constants $y_1^*,\ldots,y_p^* > 0$ such that in $Q = \{y\in\mathbb{R}^p: y_j> -y_j^*\}$:

$$\partial g_j/\partial y_{j-1}>0 \ (2\leq j\leq p), \ \partial g_1/\partial y_p< 0, \tag{6.16}$$

$$\partial g_j/\partial y_j<0 \ (2\leq j\leq p) \tag{6.17}$$

$$g_j(-y^*_{j-1},-y^*_j)\geq 0,\quad g_1(y_p,-y^*_1)>0\ (y_p\geq -y^*_p), \tag{6.18}$$

$$g_1(y_p,y_1)<0\quad (y_1>0,\ y_p>0),\ g_1(y_p,y_1)>0\ (y_1<0,y_p<0), \tag{6.19}$$

$\partial g_1/\partial y_1$ is bounded above in Q.

For such a system they proved :

Theorem 6.2. (Hastings,Tyson,Webster,1977). Let the Jacobian matrix

$$E=\left(\frac{\partial g_i}{\partial y_j}(0)\right)\quad i,j=1,2,\ldots,p$$

have at least one eigenvalue with positive real part, but no repeated eigenvalues. Then system (6.14) has a nonconstant periodic solution.

Because of the inequalities (6.16) the application of this theorem to (6.13) is restricted to the repressible case,i.e. $H_1=-1$.

In order to express the conditions (6.16)-(6.19) by a minimal number of assumptions on the functions $U_{i,i-1}$ in system (6.1) the following lemma is proved which also gives bounds for the amplitudes of oscillatory solutions.

Lemma 6.3. Assume the functions $U_{i,i-1}: \mathbb{R}\to\mathbb{R}_+$, $i=1,2,\ldots,n$, to be continuous, nonnegative and nondecreasing. Let $H_1=-1$, $H_2=\ldots=H_n=1$. Define a box $P\subset\mathbb{R}^p$ by

$$P=\left\{(y_1,\ldots,y_p):-\beta_i\leq\alpha_i y_j\leq\varkappa_i,\ \text{if } j=m+\sum_{l<i}k_l,\ 1\leq m\leq k_i\right\},$$

where $\varkappa_1=U_{1,n}(\bar{v}_p)$, $\varkappa_i=U_i(\varkappa_{i-1}/\alpha_{i-1})$, $i=2,\ldots,n$,

$\beta_1=U_1(\varkappa_n/\alpha_n)$, $\beta_i=-U_i(\beta_{i-1}/\alpha_{i-1})$, $i=2,\ldots,n$.

Then every trajectory $y(t)=(y_1(t),\ldots,y_p(t))$, $t\geq 0$, solving system (6.13) approaches P as $t \to \infty$, moreover

$$y(0) \in P \text{ implies } y(t) \in P \quad \text{for all } t > 0.$$

Proof. It follows from the definition of U_1 that $U_1(\xi) \geq -\varkappa_1$ for all real ξ .Hence $\dot{y}_1 \leq \varkappa_1 - \alpha_1 y_1$. If $y_1 > \varkappa_1/\alpha_1$ then $\dot{y}_1 < 0$. Therefore $\limsup y_1(t) = \varkappa_1/\alpha_1$ as $t \to \infty$, and $y_1(0) \leq \varkappa_1/\alpha_1$ implies $y_1(t) \leq \varkappa_1/\alpha_1$ for $t > 0$.
It is easily shown that if the conditions

$$\limsup y_j(t) \leq \varkappa_i/\alpha_i \text{ as } t \to \infty, \tag{6.20}$$

and

$$y_j(0) \leq \varkappa_i/\alpha_i \text{ implies } y_j(t) \leq \varkappa_i/\alpha_i \quad \text{for all } t \geq 0 \tag{6.21}$$

are satisfied for $j = j_o < p$ then they also hold for $j = j_o + 1$. Having proved in this way that (6.20) and (6.21) hold for $j = 1, \ldots, p$, it is possible to show that

$$\liminf y_j(t) = -\beta_i/\alpha_i \text{ as } t \to \infty \tag{6.22}$$

and that

$$y_j(0) = -\beta_i/\alpha_i \text{ implies } y_j(t) = -\beta_i/\alpha_i, t > 0, \tag{6.23}$$

is true for $j = 1$.
By induction it follows that (6.22) and (6.23) hold also for $j = 2, \ldots, p$, Q.E.D.

It is now possible to derive the following result as a consequence of theorem 6.2.

Theorem 6.4. The system (6.1) of integral equations has a non-constant periodic solution $v(t) = (v_1(t), \ldots, v_n(t))$ if the following conditions are satisfied:

(i) the input $(E_1,\dots,E_n)$ is independent of time,

(ii) the absolute value of the kernel functions $h_{i,i-1}$ is given by (6.6), and their signs $H_{i,i-1}=\text{sign } h_{i,i-1}$ obey

$$\prod_{i=1}^{n} H_{i,i-1} = -1,$$

(iii) the functions $U_{i,i-1}: \mathbb{R} \longrightarrow \mathbb{R}_+$ are continuously differentiable and strictly increasing, $i=1,2,\dots,n$,

(iv) the characteristic equation (5.46) associated with the steady state solution $\bar{v}=(\bar{v}_1,\dots,\bar{v}_n)$ of system (6.1) has only simple roots, one of which has a positive real part.

Proof. First the system (6.1) is transformed into the equivalent system (6.3). Because of lemma 6.1 the coupling terms H_i can be assumed to equal +1 except $H_1=-1$. Then the integral system (6.3) is transformed into the differential system (6.13), which has the form (6.14). The conditions (6.16), (6.17), and (6.19) are easily verified. Note that because of lemma 6.3 they have only to hold in the set P instead of the set Q. For (6.18) choose $y_1^*=\beta_1/\alpha_1+\varepsilon$, $\varepsilon>0$ small, $y_j^*=\beta_i/\alpha_i$, $2\leq j\leq p$, where β_i is defined in lemma 6.3 and the index relation $j=m+\sum_{l<i} k_l$, $1\leq m\leq k_i$, holds. Then the application of theorem 6.2 yields a periodic solution of system (6.13). According to (6.10) and (6.12) the variables $y_j(t)$ of this solution with $j=1+\sum_{l<i} k_l$ have to be identified with $w_i(t)$, $i=1,2,\dots,n$, respectively. These variables satisfy the equations (6.3), Q.E.D.

Remark 1. Condition (iv) of the theorem means in particular that $\bar{v}$ is unstable. For n=1 (single neuron) a necessary and sufficient condition that a root of (5.46) has positive real part is given in theorem 5.7. For n=2 the corresponding condition is contained in (iv) of sect. 5.5. The generalization to arbitrary dimension n is straightforward. In particular the arguments preceding theorem 5.11 show that the steady state is most easily destabilized (i.e. a root with positive real part occurs) if all decay rates α_i are equal, $\alpha_1=\alpha_2=\ldots=\alpha_n$. In this case the roots of (5.46) are simple and the condition (iii) of theorem 5.12 is necessary and sufficient for the existence of a root with positive real part. It can be concluded that sustained oscillations are more likely the larger the sum $k=\sum_{i=1}^{n} k_i$ is, since the minimal value $|U'|=\sum_{i=1}^{n} U'_{i,i-1}(\bar{v}_{i-1})$ necessary for destabilization decreases with increasing k.

The proof of the theorem of Hastings, Tyson and Webster allows one to say something about the phase relationships between the components $(v_1(t),\ldots,v_n(t))$ of oscillatory solutions v(t) of system (6.1) in the repressible case ($\prod_{i=1} H_i=-1$).
Consider again the corresponding differential equation system (6.13). As shown in lemma 6.3 every solution of (6.13) (with $H_1=-1$) ultimately enters the box P, which is positively invariant with respect to the flow. In particular the trajectory of the periodic solution is contained in P. The box P is now divided into 2^p subboxes P_k, $k=0,1,\ldots,2^p-1$:
If the integer k has the binary expansion

$$k=a_1(k)\ldots a_p(k),\quad a_j(k)\in\{0,1\}\ ,$$

define

$$P_k=\{y\in P:(-1)^{a_j(k)}\cdot y_j\leq 0,\ 1\leq j\leq p\}\ .$$

The j-th component of $y\in P_k$ is nonpositive if the j-th digit is zero, and it is nonnegative if the j-th digit of k is 1.
Consider the subset

$$V=P_{00\ldots0}\cup P_{100\ldots0}\cup P_{110\ldots0}\cup\ldots\cup P_{11\ldots1}\cup P_{011\ldots1}\cup\ldots\cup P_{00\ldots01} \tag{6.24}$$

of P containing just those $y\in P$ which have in the sequence

$y_1, \ldots, y_p$ at most one strict sign change.
Hastings, Tyson and Webster show that V is positively invariant with respect to the flow of equ. 6.13 (with $H_1=-1$) and that on the assumptions (i)-(iv) of theorem 6.4 any solution $y(t)$ of (6.13) with $y(0) \in V-\{0\}$ passes again and again through the $P_k \subset V$ in the order in which they are listed in the definition of V (in particular the periodic trajectory of theorem 6.4 is contained in V). By help of lemma 6.1 it is possible to generalize this result to arbitrary repressible coupling types. The method will become sufficiently clear by an example.

Example 6.5. Given a closed chain of three inhibitory neurons with weight functions h_i as in (6.6), $k_1=k_2=k_3=2$. Then $H_1=H_2=H_3=-1$.
The situation described above is obtained by transforming the perturbed variables $w_i=v_i-\bar{v}_i$ of the potentials v_i according to the transformation (6.5):

$$\tilde{w}_1=w_1, \quad \tilde{w}_2=-w_2, \quad \tilde{w}_3=w_3 \quad . \tag{6.25}$$

The corresponding differential system (6.13) consists of $p=k_1+k_2+k_3=6$ equations with variables $y_1, \ldots, y_6$, obeying

$$\tilde{w}_1=y_{k_1}=y_2, \quad \tilde{w}_2=y_{k_1+k_2}=y_4, \quad \tilde{w}_3=y_{k_1+k_2+k_3}=y_6 . \tag{6.26}$$

If the conditions of theorem 6.4 hold for this chain, then any solution vector $y(t)=(y_1(t), \ldots, y_n(t))$ of system (6.13) starting in P_{000000} moves successively through the 2p=12 sets P in (6.24) from left to right infinitely often.
According to 6.26 the simultaneous signs of $(\tilde{w}_1, \tilde{w}_2, \tilde{w}_3)$ are successively

$\tilde{w}_1$	-	-	+	+	+	+	+	+	-	-	-	-
$\tilde{w}_2$	-	-	-	-	+	+	+	+	+	+	-	-
$\tilde{w}_3$	-	-	-	-	-	-	+	+	+	+	+	+

Because of (6.25) the sign pattern of (w_1, w_2, w_3) during one cycle is

w_1	-	+	+	+	-	-
w_2	+	+	-	-	-	+
w_3	-	-	-	+	+	+

Obviously this qualitative description of the phase relations supplies a good test for the validity ofthe model. Analogously the sign pattern of the n oscillating components of arbitrary closed chains of length n is easily obtained. It is possible that these considerations can be applied to the locomotory system of the leech.

6.3 Oscillations in systems with delays

The stability investigations in sects.5.4 and 5.5 showed that delays in the responses of the cells to their own activity or to that of other cells can destabilize a steady state, which would be stable if the delay were not present. Therefore it is not surprising that delays can also become a source of undamped oscillations. This feature has long been recognized in the theory of differential and functional-differential equations and in various areas of applied sciences. In biological systems delays are of extreme importance, since their complexity makes delays an inevitable concomitant. A (nearly) up-to-date monograph on the mathematical theory of systems with delays is (Hale,1977), many biological applications and further mathematical results are contained in the books (Cushing, 1977) and (MacDonald, 1978).
Most mathematical results on periodic oscillations in nonlinear systems with discrete delays concern first order systems, the most general form so far investigated being

$$\dot{x}(t)=f(x(t), x(t-\tau)), \tag{6.27}$$

x a scalar variable, f a continuous function of two variables, and $\tau > 0$ the delay assumed to be constant (Kaplan and Yorke, 1977; Hadeler, 1979). The most successful method was clarified by Nussbaum,1974,1973. In principle this method, which applies a theorem of Browder on ejective fixed points to a cone in the state space, can be used for arbitrarily high dimensional systems, and indeed it will help subsequently to derive a result (theorem 6.7) on a second order system.

Consider first a single self-inhibitory cell with a temporal weight function as in equ. (5.23) :

$$h_a(t) = -\alpha\, H(t-\tau)\ \exp\,(-\alpha(t-\tau)), \tag{6.28}$$

where H denotes Heaviside's function $(H(t)=0\,(t<0), =1\,(t\geq 0))$. Assume the cell to have a temporally constant input $E_1(t)=\bar{E}_1$. Then according to the results in sect. 4.2 a) the system (5.1) with n=1 has a unique stationary solution $\bar{v}_1$. With the abbreviation

$$U_1(w_1)=U_{11}(w_1+\bar{v}_1)-U_{11}(\bar{v}_1) \tag{6.29}$$

the perturbed variable $w_1(t)=v_1(t)-\bar{v}_1$ satisfies (see (5.2))

$$w_1(t) = -\alpha \int_{-\infty}^{t-\tau} U_1(w_1(t'))\ \exp(-\alpha(t-t'-\tau))dt' . \tag{6.30}$$

Differentiating both sides of this equation the following differential-difference equation is obtained:

$$\dot{w}_1(t) = -\alpha\, U_1(w_1(t-\tau)) - \alpha\, w_1(t) . \tag{6.31}$$

Periodic solutions of this type of equations seem to have been investigated first by Chow, 1974, as a model for blood cell production, and by Pesin, 1974. Later on Coleman and Renninger, 1976, arrived at this equation in the context of a homogeneous lateral inhibitory network with applications to the Limulus eye.

They concentrated on the nonlinearity $U_{11}(v)=\max(0,v)$ and found explicit solutions. With the same application in mind, Hadeler and Tomiuk, 1977, gave a more general result on the existence of periodic solutions. Their theorem is restated here in form of assumptions on α, U_1 and τ .

<u>Theorem 6.6.</u>(Hadeler, Tomiuk,1977). Let U_1: $\mathbb{R} \longrightarrow \mathbb{R}$ be a continuous function and $\alpha > 0$ a real number satisfying

(i) $\xi U_1(\xi) > 0 \quad$ if $\xi \neq 0$,

(ii) U_1 is bounded below, i.e. there is $\varkappa > 0$ such that

$$U_1(\xi) = -\varkappa \quad \text{for all } \xi \quad ,$$

(iii) U_1 is differentiable at 0,

(iv) $U_1'(0)$ obeys the inequality (5.29) with $>$ instead of $<$.

Then equ.(6.31) has a nonconstant periodic solution with a period greater than 2τ .

The proof of this theorem is based on the method of Nussbaum. The authors use as an invariant cone the set of all continuous functions $\varphi : [-\tau, 0] \to \mathbb{R}$ obeying $\varphi(-\tau)=0$ and $e^{\alpha t}\varphi(t)$ nondecreasing, $-\tau \leq t \leq 0$.

The result of theorem 6.6 will be extended below to the case that the neuron has a weight function h_b as defined in (5.24):

$$h_b(t) = -\alpha^2 H(t-\tau) \cdot \exp(-\alpha(t-\tau)), \; t \geq 0, \alpha > 0. \tag{6.32}$$

Then with (6.29) the perturbed variable $w_1 = v_1 - \bar{v}_1$ satisfies

$$w_1(t) = \int_{-\infty}^{t-\tau} U_1(w_1(t'))h_b(t-t')dt' \quad . \tag{6.33}$$

Introducing the auxiliary variable

$$z(t) = -\alpha \int_{-\infty}^{t-\tau} U_1(w_1(t')) \exp(-\alpha(t-t'-\tau))dt', \tag{6.34}$$

the differential equation system corresponding to the integral equation (6.33) is

$$\begin{aligned} \dot{w}_1(t) &= \alpha z(t) - \alpha w_1(t) \\ \dot{z}(t) &= -\alpha U_1(w_1(t-\tau)) - \alpha z(t). \end{aligned} \tag{6.35}$$

It will turn out that this system can be treated together with the case of a pair of excitatory-inhibitory coupled neurons with weighting functions as in (5.28):

$$h_1(t)=h_{12}(t)=\alpha_1\ H(t-\tau_1)\ \exp(-\alpha_1(t-\tau_1)) \tag{6.36a}$$

$$h_2(t)=h_{21}(t)=-\alpha_2 H(t-\tau_2)\ \exp(-\alpha_2(t-\tau_2)), \tag{6.36b}$$

where $\alpha_1, \alpha_2 > 0$, $\tau_1, \tau_2 \geqq 0$ are constants, $h_{ii} \equiv 0$ (no self-interactions!).
Again the input $E=(E_1,E_2)$ to the pair is assumed to be constant, $E=\bar{E}$. Following sect. 4.2b)

$$\begin{aligned} U_2(w_2)&=U_{12}(w_2+\bar{v}_2)-U_{12}(\bar{v}_2) \quad , \\ U_1(w_1)&=U_{21}(w_1+\bar{v}_1)-U_{21}(\bar{v}_1) \end{aligned} \tag{6.37}$$

the perturbation variables $w_1=v_1-\bar{v}_1$ and $w_2=v_2-\bar{v}_2$ obey

$$\begin{aligned} \dot{w}_1(t)&=\alpha_1 U_2(w_2(t-\tau_1))-\alpha_1 w_1(t), \\ \dot{w}_2(t)&=-\alpha_2 U_1(w_1(t-\tau_2))-\alpha_2 w_2(t) \ . \end{aligned} \tag{6.38}$$

These equations can be simplified by shifting the time scale for one of the two dependent variables:
Let $\tau=\tau_1+\tau_2$ and $z(t)=w_2(t-\tau_1)$. Then (6.38) is equivalent to

$$\begin{aligned} \dot{w}_1(t)&=\alpha_1 U_2(z(t))-\alpha_1 w_1(t) \\ \dot{z}(t)&=-\alpha_2 U_1(w_1(t-\tau))-\alpha_2 z(t). \end{aligned} \tag{6.39}$$

<u>Theorem 6.7.</u> The system (6.38) of differential-difference equations has a nonconstant periodic solution with a period larger than 2τ if the following conditions hold:

(i) the continuous functions $U_1: \mathbb{R} \to \mathbb{R}$, $U_2: \mathbb{R} \to \mathbb{R}$ satisfy

$$\xi U_i(\xi) > 0 \text{ for all } \xi \neq 0, \ i=1,2,$$

(ii) at least one of the two functions is bounded below, i.e.

$$U_i(\xi) \geq -\varkappa \quad \text{for } i=1 \text{ or } i=2 \text{ and some } \varkappa > 0,$$

(iii) both functions are differentiable at $\xi = 0$,

(iv) the product $U' = U_1'(0) \cdot U_2'(0)$ satisfies the lower inequality in corollary 5.11 with $\tau = \tau_1 + \tau_2$.

Proof (covering the following 9 pages).
Changing the time scale by a factor $1/\tau$ the system (6.39) can be cast into the more convenient form

$$\dot{x}(t) = g(y(t)) - ax(t) \qquad (6.40a)$$
$$\dot{y}(t) = -f(x(t-1)) - by(t) \qquad (6.40b)$$

with a delay $\tau = 1$. It is supposed that the conditions (i)-(iv) of the theorem hold for the functions g and f respectively. The first lemma will show that from an assumption weaker than condition (iv) the solutions of (6.40) are oscillatory if the initial conditions are elements of the cone $K \subset C([-1,0]) \times \mathbb{R}$ defined by

$$K = \{\psi = (\varphi, y_0) : \varphi(-1) = 0, e^{at}\varphi(t) \text{ nondecreasing on } [-1,0], y_0 \geq 0\} .$$

Lemma 6.8. If $f'(0)g'(0) > ab/(e^{\min(a,b)} - 1)$, then the solution $S(t) = (x(t), y(t))$, $t \geq 0$, of system (6.40) corresponding to an initial condition $\psi \in K - \{0\}$ satisfies:

(i) the zeros of $x(t)$ for $t > 0$ form an infinite sequence (z_k), $k=1,2,\ldots$ with $x(z_k) = 0$, $z_{k+1} - z_k > 1$
$\dot{x}(z_{2k-1}) < 0$, $\dot{x}(z_{2k}) > 0$, $y(z_{2k-1}) < 0$, $y(z_{2k}) > 0$,
$y(z_{2k-1}+1) < 0$, $y(z_{2k}+1) > 0$,

(ii) the function $y(t)\exp(at)$ is nondecreasing on the interval $(z_{2k}, z_{2k}+1)$ and nonincreasing on the interval $(z_{2k-1}, z_{2k-1}+1)$,

(iii) the map $\mathcal{F}: K \to K$, defined by $\mathcal{F}(0)=0$ and, for $\psi \neq 0$, $\mathcal{F}(\psi)=(\tilde{\varphi}, \tilde{y}_0)$, where $\tilde{\varphi}(t)=x(t+z_2+1)$, $\tilde{y}_0=y(z_2+1)$, has the following continuity property:
For each $M>0$ there is $\tilde{M}>0$ such that $\|\psi\| \leq M$ implies $\|\mathcal{F}(\psi)\| \leq \tilde{M}$; $\tilde{M} \to 0$ as $M \to 0$.

Proof of lemma 6.8. The idea is to follow the trajectory $S(t)$ along one revolution in the (x,y)-plane. Let $M>0$ and assume $\|\psi\| = \max\,(y_0, \sup_{t\in[-1,0]} \varphi(t)) \leq M$. Let $\mathbb{R}^2_+ = \{(x,y): x \geq 0, y>0\}$. Define $t_1 = \inf\{t \geq 0: S(t) \notin \mathbb{R}^2_+\}$. As long as $S(t) \in \mathbb{R}^2_+$ either $S(t)$ is situated to the left or to the right of the curve $L_1 = \{(x,y) \in \mathbb{R}^2_+ : x=g(y)/a\}$. If $S(t)$ is strictly to the left of L_1 then $\dot{x}(t)>0$ (because of equ. (6.40a)). Therefore $x(t)>0$ for $0<t<t_1$. Since moreover $\varphi \geq 0$ it follows from (6.40b) that

$$\dot{y}(t) < -by(t),$$

hence

$$y(t) \leq y_0 \exp(-bt) \text{ for } 0<t<t_1 . \tag{6.41}$$

If $S(t)$ is to the right of L_1 then $\dot{x}(t) \leq 0$, therefore

$$x(t) \leq \max\,(\varphi(0), \max_{0 \leq \xi \leq M} g(\xi)/a) = M_1 \text{ for } 0 \leq t < t_1 . \tag{6.42}$$

There are constants $c, d, \varepsilon, \delta > 0$ such that

$$0<c<f'(0),\ 0<d<g'(0)/a<(1+\varepsilon)d,\ cd/(1+\varepsilon)>b/(e^{\min(a,b)}-1), \tag{6.43}$$

$$d|y|<|g(y)/a|<(1+\varepsilon)d|y| \quad (0<|y| \leq \delta),$$

$$c|x|<|f(x)| < (1+\varepsilon)c|x| \quad (0<|x| \leq \delta).$$

It will be shown that t_1 is finite. Assume $t_1 = \infty$.
The estimate (6.41) implies $g(y(t)) \to 0$ as $t \to \infty$, and hence with (6.40a) $\lim x(t)=0$ as $t \to \infty$. Choose $\tau_1 \geq 0$ so large that $y(t) \leq \delta$ and $x(t-1) \leq \delta$ for all $t \geq \tau_1$. There is $\tau_2 \geq \tau_1$ such that $x(t-1) \geq dy(t-1)$ for all $t \geq \tau_2$ (this follows from $\lim y(t)=0$ as $t \to \infty$ and $\dot{x}(t) > 0$ as long as $x(t) \leq dy(t) < g(y(t))/a$).
For $t \in [\tau_2-1, \tau_2]$ either $x(t) \leq g(y(t))/a$ or $x(t) > g(y(t))/a$.
In the first case, since $y(t')$ decreases and $\dot{x}(t') < 0$ if $x(t') > (1+\varepsilon)d\, y(t')$,

$$x(\tau_2) \leq (1+\varepsilon)dy(t) \leq (1+\varepsilon)x(t) \tag{6.44}$$

In the second case $\dot{x}(t) < 0$, therefore $y(t) > x(\tilde{t})$, where $\tilde{t} < t$ is defined as $\tilde{t} = \sup\{t' : t' = \tau_2 \text{ or } x(t') \leq g(y(t'))/a\}$.
Hence with (6.44)

$$x(t) \geq x(\tau_2)/(1+\varepsilon) \quad \text{for } t \in [\tau_2-1, \tau_2] . \tag{6.45}$$

Integration of (6.40b) leads to

$$y(t) = -\int_{t_o}^{t} f(x(t'-1))e^{-b(t-t')}dt' + y(t_o)e^{-b(t-t_o)} . \tag{6.46}$$

Taking $t_o = \tau_2$ and $t = \tau_2+1$ it follows from the estimates above that

$$y(\tau_2+1) \leq -c\int_{\tau_2}^{\tau_2+1} x(\tau_2)(1+\varepsilon)^{-1}e^{-b(t-t')}dt' + x(\tau_2)e^{-b}/d$$

$$= -c\, x(\tau_2)(1-e^{-b})/[(1+\varepsilon)b] + x(\tau_2)e^{-b}/d .$$

Because of (6.43) the last expression is negative, and thus a contradiction to $t_1=\infty$ is obtained.
Obviously $S(t)$ can leave $\mathbb{R}^2_+$ only across the x-axis. Therefore $y(t_1)=0$ and $x(t_1) \geq 0$. Indeed $x(t_1) > 0$ as follows from the

integral representation of $x(t)$,

$$x(t)=\int_{t_o}^{t} g(y(t'))e^{-a(t-t')}dt' + x(t_o)e^{-a(t-t_o)} \qquad (6.47)$$

with $t_o=0$ and $t=t_1$.

Let $t_2=\inf\{t\geq t_2 : y(t)\neq 0\}$. Equation (6.46) with $t_o=t_1$ implies that $t_2\leq t_1+1$, since $x(t)\geq 0$ $(-1\leq t\leq t_1)$ and $x(t)>0$ immediately before the time t_1. Moreover $y(t)<0$ for $t>t_2$ as long as $x(t-1)\geq 0$.

Because of (6.47) with $t_o=t_1$ the function $x(t)$ decreases in the interval $[t_1,t_2]$,and $x(t_2)>0$.

Let $z_1>t_2$ denote the minimal positive number with $x(z_1)=0$ (if $x(t)$ has no positive zero, $z_1=\infty$). It follows from (6.40a) that $x(t)$ decreases in the interval $[t_2,z_1)$, moreover from (6.47)

$$x(t)\leq x(t_2)\exp(-a(t-t_2)), \qquad t_2\leq t<z_1.$$

Assume $z_1=\infty$. Then $\lim x(t)=0$ as $t\to\infty$. Hence with $t_o=t_2+1$ in (6.46) $\lim y(t)=0$ as $t\to\infty$. Choose $\tau_3\geq t_2+1$ so large that $y(t)\geq-\delta$ and $x(t-1)\leq\delta$ for $t\geq\tau_3$.

Consider the line $L_2=\{(x,y):0\leq x,\ by=-cx\}$. There is $\tau_4\geq\tau_3$ such that $S(t)$ below L_2, i.e. $by(t)\leq -cx(t)$, for all $t\geq\tau_4$ (since $x(t)\to 0$ as $t\to\infty$ and $\dot{y}(t)\leq -cx(t-1)-by(t)\leq -cx(t)-by(t)<0$ as long as $S(t)$ strictly above L_2).

Let $t\in[\tau_4,\tau_4+1]$. With the estimate $f(x)>cx$ for $x>0$ the integral representation (6.46) implies

$$y(t)\leq -c\int_{\tau_4}^{t} x(t'-1)e^{-b(t-t')}dt'+y(\tau_4)e^{-b(t-\tau_4)}.$$

Since $\dot{x}(t'-1)\leq 0$ and $y(\tau_4)\leq -cx(\tau_4)/b$, it follows

$$y(t)\leq -cx(\tau_4)\ (1-e^{-b(t-\tau_4)})/b-cx(\tau_4)\cdot e^{-b(t-\tau_4)}/b,$$

$$y(t)\leq -cx(\tau_4)/b\quad .$$

The last inequality is used in (6.47) with $t_o=\tau_4$ and $t=\tau_4+1$ to obtain

$$x(\tau_4+1)\leq -dcx(\tau_4)\ (1-e^{-a})/b+x(\tau_4)e^{-a}<0.$$

This result contradicts the assumption $z_1=\infty$ .Therefore z_1 is finitc.
From equ. (6.46) with $t_o=t_1$ and from $0\leq x(t)\leq \max(M,M_1)$, $t\in[-1,z_1]$ it follows that

$$y(t)\geq -\max\left\{f(\xi):0\leq\xi\leq\max(M,M_1)\right\}/b=-M_2,\quad t\in[t_1,z_1]\ ,$$

and $y(z_1)<0$. Equation (6.40a) implies $\dot{x}(z_1)<0$.

The function $x(t)\exp(at)$ is nonincreasing in $[z_1,z_1+1]$, since according to equ. (6.47)

$$\exp(at)x(t)=\int_{z_1}^{t} g(y(t'))\ \exp(at')dt',$$

where the integrand is negative (as can be seen from (6.46) with $t_o=z_1$).
Obviously $y(z_1+1)<0$.
From the integral representations of $y(t)$ and $x(t)$ the following bounds can be obtained. For $t\in[z_1,z_1+1]$

$$y(t)\geq y(z_1)-\int_{z_1}^{z_1+1} bM_2\exp(-b(z_1+1-t'))dt'\geq -M_2(2-e^{-b})=-M_3\quad ,$$

$$x(t)\geq\int_{z_1}^{z_1+1}\min\left\{g(\xi):-M_3\leq\xi\leq 0\right\}\exp(-a(z_1+1-t'))dt'=-M_4\ .$$

The constant $\overline{M}=\max\{M,M_1,\ldots,M_4\}$ is independent of $x(t)$ and $y(t)$ and satisfies $\overline{M} \to 0$ as $M \to 0$.
Take now as a new initial condition of the system (6.40) the pair $\overline{\Psi}=(\overline{\varphi},\overline{y}_0)$ defined by

$$\overline{\varphi}(t)=x(z_1+1+t),t\in[-1,0]\ ,\overline{y}_0=y(z_1+1)\quad .$$

Then $\overline{\Psi}\in -K$ and $0<\|\overline{\Psi}\|\leq \overline{M}$. Because of the symmetry properties of the system (6.40) the same reasoning as applied to the initial value problem Ψ shows that the solution $\overline{S}(t)=(\overline{x}(t),\overline{y}(t))$, corresponding to $\overline{\Psi}$ obeys the following conditions: there is a first zero $\overline{z}_1$ of $\overline{x}(t)$, $t>0$, $\overline{y}(\overline{z}_1+1)>0$, $\overline{x}(t)\exp(at)$ is nondecreasing on $[\overline{z}_1,\overline{z}_1+1]$, and there are bounds $\tilde{M}_1,\ldots,\tilde{M}_4$ related to $\overline{M}$ in the same way as $M_1,\ldots,M_4$ to M. By the choice of $\overline{\Psi}$ the number $z_2=z_1+1+\overline{z}_1$ is the second zero of $x(t)$, and $\tilde{\varphi}(t)=x(t+z_2+1)=\overline{x}(t+\overline{z}_1+1)$, $\tilde{y}_0=y(z_2+1)=\overline{y}(\overline{z}_1+1)$ have the properties stated in the lemma. Repetition of these arguments with respect to $\varphi(t)=x(t+z_k+1)$, $y_0=y(z_k+1)$, $k=2,3,\ldots$, completes the proof of lemma 6.8.

<u>Lemma 6.9.</u> Assume the hypothesis of lemma 6.8.Then the map $\mathcal{F}$ is continuous and compact on K.

<u>Proof.</u> It follows from lemma 6.8 (iii) that $\mathcal{F}$ is continuous at 0. The map $\Psi \to z_2(\Psi)$ is continuous on $K-\{0\}$, since the solutions of (6.40) depend continuously on the initial conditions. Therefore $\mathcal{F}$ is continuous on $K-\{0\}$.To prove the compactness let A be a bounded subset of K, say $\|\psi\|<M,\psi\in A$. Because of lemma 6.8 (iii) the range $\mathcal{F}(A)$ is bounded. Moreover it was shown in the proof of lemma 6.8 that the solutions $S(\psi)=(x(\psi),y(\psi))$ corresponding to initial conditions $\psi\in A$ are uniformly bounded on $[-1,z_2(\psi)+1]$ (namely by the constant $\tilde{M}$).Therefore the components $\tilde{\varphi}(\psi)$ of $\mathcal{F}(\psi)$,as obtained by a shift of the solution of equ.(6.40a),are equicontinuous for all $\psi\in A$.The theorem of Arzela-Ascoli implies that $\mathcal{F}(A)$ has compact closure,Q.E.D.

Lemma 6.10. Assume the hypothesis of lemma 6.8. Let $f(\xi)\geq -\varkappa_1$ for all ξ or $g(\xi)\geq -\varkappa_2$ for all ξ. Then the operator $\mathcal{F}$ maps the closed bounded, and convex set $D=\{(\varphi,y_0)\in K: \|\varphi\|\leq \varkappa^*(1-e^{-a})/a, y_0\leq \varkappa_1/b\}$ into itself, where $\varkappa^*=\max\{g(\xi): 0\leq \xi\leq \varkappa_1/b\}$, (if f is not bounded below, then $\varkappa_1=\max\{-f(\xi): -\varkappa_2/a\leq \xi\leq 0\}$).

Proof. Let $\psi=(\varphi,y_0)\in D$ and $\mathcal{F}(\varphi,y_0)=(\tilde{\varphi},\tilde{y}_0)$.
Case 1: Assume f is bounded below by $-\varkappa_1$. It follows from lemma 6.8 that $y(z_1+1)<0$. Equ. (6.46) with $t_0=z_1+1$ implies

$$y(t)=-\int_{z_1+1}^{t} f(x(t'-1))e^{-b(t-t')}dt'\leq \varkappa_1/b \text{ for all } t\geq z_1+1 \quad .$$

In particular $\tilde{y}_0=y(z_2+1)\leq \varkappa_1/b$.

Let $\varkappa^*$ be defined as above. Then equ. (6.47) implies for $t\in[z_2,z_2+1]$ that

$$\tilde{\varphi}(t-z_1-1)=x(t)=\int_{z_2}^{t} g(y(t'))e^{-a(t-t')}dt'\leq \varkappa^*(1-e^{-a})/a \quad .$$

Case 2: Assume g is bounded below by $-\varkappa_2$ and f is not bounded below. Then (6.40a) implies $\dot{x}(t)\geq -\varkappa_2-ax(t)$, and hence $\dot{x}(t)>0$ if $x(t)<-\varkappa_2/a$. Therefore $x(t)\geq -\varkappa_2/a$ for all $t\geq -1$, which implies

$$f(x(t-1))\geq -\varkappa_1=-\max\{|f(\xi)|: -\varkappa_2/a\leq \xi\leq 0\} \text{ for all } t\geq 0 \quad .$$

Now the arguments of case 1 apply, Q.E.D.

The map $\mathcal{F}$ is constructed in such a way that each of its fixed points represents an initial condition corresponding to a periodic solution of system (6.40). However, in order to find a nonconstant periodic solution, it has to be shown that $\mathcal{F}$ has a fixed point which is different from zero. This will be achieved by demonstrating that zero is an ejective point of $\mathcal{F}$.

Then lemma 6.12, a "nonejective fixed point theorem" ensures the existence of a nontrivial periodic solution.

Definition 6.11. Let X be a topological space, $x_o \in X$, and U an open neighborhood of x_o. Let $F: U - \{x_o\} \to X$ be continuous. Then x_o is called an ejective point of F if (and only if) there is a neighborhood U_o of x_o such that for each $x \in U_o - \{x_o\}$ there is an integer $i = i(x)$ satisfying $F^i(x) \notin U_o$.

Lemma 6.12. (Browder, 1965). Let D be a closed, bounded, convex set of infinite dimension in a Banach space, and let $F: D \to D$ be a continuous and compact map. Then F has a fixed point which is not ejective.

In order to prove that O is ejective the following lemma is needed, stating some equivalences of instability.

With the abbreviations $\alpha = a+b$, $\beta = ab$, $\gamma = f'(0) \cdot g'(0)$, the characteristic equation associated with system (6.40) is

$$\lambda^2 + \alpha\lambda + \beta + \gamma e^{-\lambda} = 0. \tag{6.48}$$

Lemma 6.13. Let α, β, γ be positive constants.

If $\alpha^2 \geq 2\beta$, then the following three conditions are equivalent.

1) Equation (6.48) has at least one solution with positive real part.
2) Equation (6.48) has precisely one solution λ with $\mathrm{Re}\,\lambda > 0$ and $0 < \mathrm{Im}\,\lambda < \pi$.
3) $\gamma > \alpha v_1/\sin v_1$, where $0 < v_1 < \pi$ and $\mathrm{ctg}\, v_1 = (v_1 - \beta/v_1)/\alpha$.

A proof of this lemma is given in (an der Heiden, 1979).

Lemma 6.14. On the assumptions of theorem 6.7, zero is an ejective point of $\mathcal{F}$.

Proof. According to a theorem of J. Hale and S.Chow (see Hale, 1977, theorem 11.2.3) O is an ejective point of $\mathcal{F}: K \to K$ if

(i) equ. (6.48) has a root λ with positive real part,

(ii) $\inf\{\|\pi_\lambda \psi\| : \psi \in K, \|\psi\| = 1\} > 0$, where $\pi_\lambda : C([-1,0]) \times \mathbb{R} \to P_\lambda$ denotes the projection operator from the state space onto the eigenspace P_λ of λ which is invariant under the solution operator of the linearized system

$$\dot{x}(t) = g'(0)y(t) - ax(t)$$
$$\dot{y}(t) = - f'(0)x(t-1)-by(t), \tag{6.49}$$

and

(iii) the function $z_2:D-\{0\} \longrightarrow [0,\infty)$, where $z_2(\psi)$ equals the second zero of the solution component $x(t)$ corresponding to ψ , is continuous and compact.

According to lemma 2.13 there is precisely one eigenvalue λ with Re $\lambda > 0$ and

$$0 < \text{Im}\,\lambda < \pi \quad . \tag{6.50}$$

The formal adjoint equation of system (6.49) is

$$\dot{u}(s) = a\,u(s) + f'(0)\,v(s+1)$$

$$\dot{v}(s) = -g'(0)\,u(s)+b\,v(s).$$

For $\phi = (\beta,\varphi) \in \mathbb{R} \times C([0,1])$ and $\psi = (\psi,\alpha) \in C([-1,0]) \times \mathbb{R}$ the associated bilinear form is

$$(\phi,\psi) = \beta\psi(0)+\varphi(0)\alpha + \int_0^1 f'(0)\varphi(\xi)\psi(\xi-1)d\xi \quad .$$

The projection $\pi_\lambda\psi$ is obtained by

$$\pi_\lambda\psi = \mathcal{B}(\mathcal{B}^*,\psi),$$

where $\mathcal{B}$ is a basis of P_λ and $\mathcal{B}^*$ is a basis of P_λ^* , the eigen-space of the adjoint equation associated with λ , $(\mathcal{B},\mathcal{B}^*)=I$ (=identity).

A basis of P_λ^* is $\phi = (g'(0)/(a+\lambda),\ 1)e^{-\lambda s}$, where the first component has to be evaluated at s=0. With $\lambda = \mu + i\nu$ the real and imaginary parts of (ϕ,ψ) are computed to be

$$\mathrm{Re}(\Phi,\Psi)=g'(0)(a+\mu)\psi(0)/|a+\lambda|^2+\alpha+\int_0^1 f'(0)e^{-\mu\xi}\cos(\nu\xi)\psi(\xi-1)\,d\xi \quad , \tag{6.51}$$

$$\mathrm{Im}(\Phi,\Psi)=-\,g'(0)\nu\psi(0)/|a+\lambda|^2-\int_0^1 f'(0)e^{-\mu\xi}\sin(\nu\xi)\psi(\xi-1)\,d\xi \quad . \tag{6.52}$$

Since $0<\nu<\pi$, for all $\Psi\in K$ the integrand in the expression for $\mathrm{Im}(\Phi,\Psi)$ is nonnegative. Assume the condition (ii) above is false. Then there is a sequence $\Psi_n=(\psi_n,\alpha_n)\in K$ with $\|\Psi_n\| = 1$ and $\|\pi_\lambda\Psi_n\|\to 0$ as $n\to\infty$. Hence $\mathrm{Re}(\Phi,\Psi_n)\to 0$ and $\mathrm{Im}(\Phi,\Psi_n)\to 0$ as $n\to\infty$. Because of (6.52) this means $\psi_n\to 0$, and then, because of (6.51), $\alpha_n\to 0$ as $n\to\infty$ also. Therefore the contradiction $\|\Psi_n\|\to 0$ as $n\to\infty$ is obtained which proves (ii).

Finally the function z_2 in (iii) is continuous as the solutions depend continuously on their initial conditions. Moreover, as the estimates in the proof of lemma 6.8 show, z_2 is bounded on $D-\{0\}$, which completes the proof.

Theorem 6.7 follows from the lemmas 6.9, 6.10, 6.14, and 6.12, and from the fact that a nontrivial fixed point of $\mathcal{F}$ is an initial condition of a nonconstant periodic solution. This remark completes the proof of theorem 6.7.

7. Homogeneous tissues with lateral excitation or lateral inhibition

7.1 Introduction

In this chapter the analysis is started concerning the qualitative behavior of neurons which are continuously distributed in neural tissues. The model for such neural fields has been derived in sect. 1.4, the describing equations are (1.16) (= impulse representation) and (1.19 (= potential representation).

The aim of the following chapters is to exhibit some modes of behavior in neural fields which cannot be observed in a small set of neurons. These modes include traveling fronts of excitation, propagation of pulses, spatially inhomogeneous oscillations and stable patterns.

Unfortunately the mathematical investigation of the solutions to integral equations (1.19), which are nonlinear Volterra integral equations with respect to the time variable t and Hammerstein equations with respect to the space variables s_k, is not very much advanced. Therefore attention is restricted to a special class of kernel functions $h(s_k,s_j;t)$ allowing one to transform the integral equations (1.19) into partial differential equations of parabolic type. Nonlinear parabolic equations have been studied extensively during the last two decades in various areas such as population genetics, chemical reaction-diffusion processes, ecology, nerve impulse theory, morphogenesis, and epidemics. It will turn out that many models and results from these different areas find new interpretations and a systematic integration in the context of neural networks.

7.2. The model

Let us consider the idealistic situation where an array of cells is ordered along an infinitely extended line identifiable with the line $\mathbb{R}$ of real numbers. Then the system (1.19) reduces to the single equation

$$v(s,t) = \int_{\mathbb{R}} U(s,s',v(s',.)) * h(s,s',.)ds'+E, \tag{7.1}$$

$-\infty<s<\infty$, $t\geq 0$.

Here the self-inhibitory term ($\tilde{U}$) has been neglected and the external input ($\bar{U}$) combined in the symbol $E=E(s,t)$. Without loss of generality $v_0(s)=0$ (otherwise substitute $u = v-v_0$).

Assume the network to be spatially homogeneously organized. Furthermore let the direct influence of an impulse $a\delta$ of strength a at location $s = 0$ and at time $t = 0$ on the location $s\in\mathbb{R}$ at time $t>0$ be given in the form

$$h(s,t) = U(a) \frac{1}{2\sqrt{\pi Dt}} e^{-s^2/(4Dt)} e^{-\alpha t} \tag{7.2}$$

with positive constants α, D, and a function $U:\mathbb{R} \longrightarrow \mathbb{R}$. $U(a)$ is positive (negative) when the influence is excitatory (inhibitory). At a distance s from the origin of the signal the response is maximal after a time $t = t_{max}$, where t_{max} satisfies

$$0 = \frac{\partial h}{\partial t}(s,t_{max}) = U(a)\frac{1}{2\sqrt{\pi D}} h(s,t)\left(-\frac{1}{2}\frac{1}{t}+s^2\frac{1}{4D}t^{-2}-\alpha\right),$$

i.e.

$$t_{max} = -\frac{1}{4\alpha} + \sqrt{\left(\frac{1}{4\alpha}\right)^2 + \frac{1}{4\alpha D}s^2} \quad , (\alpha>0)$$

$$t_{max} = \frac{s^2}{2D} \quad , (\alpha=0).$$

Therefore the time of maximal response is an increasing

function of the distance of the signal. The larger the "diffusion constant" D the shorter the time the maximum is reached. For each fixed $t>0$ the response profile $h(s,t)$ is a Gaussian function. Hence the strength of coupling between the cells at two locations decreases with their distance.

The global effect of the signal extinguishes at a rate $\alpha>0$.

Finally it is assumed that $U(s,s',v) = U(v)$ and that the mutual couplings depend only on the distance: $h(s,s',t) = h(|s-s'|,t)$. Then, because of the relation

$$h_t = Dh_{ss} - \alpha h, \qquad (t>0),$$

the quantity

$$w(s,t) = \int_{\mathbb{R}} U(v(s',.)) * h(|s-s'|,.)ds'$$

by (7.1) obeys the parabolic equation

$$w_t + \alpha w - Dw_{ss} = U(w+E). \tag{7.3}$$

The potential v is obtained from w by

$$v = w + E. \tag{7.4}$$

On the condition $E=$ constant, (7.3) and (7.4) imply

$$v_t - D\, v_{ss} = U(v) - \alpha v + I, \tag{7.5}$$

where $I = \alpha E$.

As the simplest case of a continuously distributed neural network we shall now investigate the qualitative behavior of solutions to (7.5), where all cells receive the same constant input represented by the value I. First the notation is simplified by introducing the function

$$f(v) = U(v) - \alpha v + I. \tag{7.6}$$

Then (7.4) reads

$$v_t - D\, v_{ss} = f(v). \tag{7.7}$$

This equation is known under the name of R.A.Fisher, who originally introduced it as a selection-diffusion model in population genetics in 1937. Continual efforts during the last decades (for review see Hadeler, 1976, and Fife, 1979) brought about many analytical results, some of which will be summarized in the following.

The qualitative behavior of the potential v as a solution of (7.7) heavily depends on the source term f. If the coupling of the cells is inhibitory (U nonpositive and decreasing) then f is a monotonic decreasing function with exactly one zero.
As to excitatory couplings, assuming a nonnegative, sigmoid and bounded function U, two generic cases have to be distinguished :
a) The slope of U is everywhere smaller than α ($U'(\xi)<\alpha$). In this situation f is again a monotonic decreasing function with one and only one zero.
b) There is some point ξ_o obeying $U'(\xi_o)>\alpha$.
Then, according to the definition (7.6), in a certain range of I the function f has exactly three zeros (the corresponding values of I fill a whole interval). The shape of f is depicted in fig. 7.1.

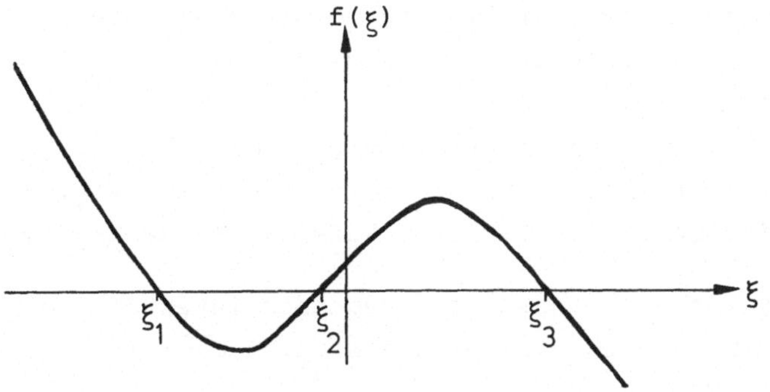

Fig. 7.1

It is particularly this case which exhibits interesting behavior.

7.3. Stationary solutions and their stability

The stationary, i.e. time-independent solutions v(s,t) = v(s) of (7.7) are determined by the ordinary differential equation

$$v''(s) + f(v(s)) = 0, \qquad -\infty < s < \infty . \tag{7.8}$$

The following discussion contains a complete determination of all solutions of (7.8).
The simplest stationary solutions are those which are also space-independent. Such constant solutions are given by the zeros of the function f. Whereas with inhibitory coupling the constant solution is unique, it may be in the case of lateral excitation that there are three constants satisfying (7.7).

The non-constant stationary solutions can be classified by help of the function

$$V(\xi) = \int_0^{\xi} f(\xi')d\xi'. \tag{7.9}$$

For each non-constant stationary solution v(s) there is a constant C such that the range $R = R(v) = \{ v(s) : -\infty < s < \infty \}$ is an interval satisfying

i) $V(\xi) < C$ if ξ is an interior point of R,

ii) the endpoints of R are $+\infty$ or $-\infty$ or satisfy $V(\xi) = C$ (for details see Fife, 1977, or Fife, 1979, section 4.3).

Conversely for each constant C and each interval R with properties i) and ii) there is a stationary solution, its range coinciding with R.

To apply this criterion observe first that as f is the derivative of V, the function V is strictly decreasing whenever f is negative and strictly increasing whenever f is positive. Therefore V has a maximum where f has a zero and negative slope (assuming f to be differentiable) and a minimum where f has a zero and positive slope.

Because of the special form (7.6) of f (U is supposed to be bounded) it follows that if f has a single zero, say at ξ_0, V has a maximum at ξ_0 and V is increasing from $-\infty$ to $V(\xi_0)$ in the interval $-\infty < \xi \le \xi_0$ and decreasing from $V(\xi_0)$ to $-\infty$ in the interval $\xi_0 \le \xi < \infty$, as indicated in fig. 7.2a)

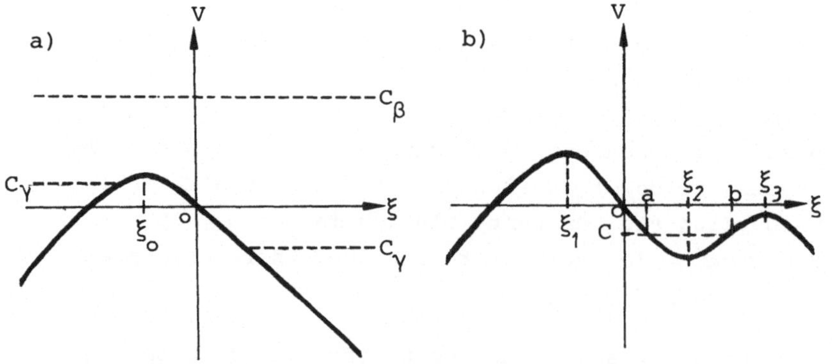

Fig. 7.2

Summarizing and also using some other observations in the proof of Fife's result the following lemma is obtained.

<u>Lemma 7.1.</u> Given a constant I and a bounded continuous function U in the expression (7.6) for the function f, assume that f has only one zero (as is generally the case for inhibitory coupling, and in the case of excitatory coupling if U has everywhere a slope smaller than α). Then there are three types of stationary solutions $v(s,t) = $ $= v(s)$ of equation (7.7):

α) exactly one constant solution,

β) infinitely many monotone solutions with range from $-\infty$ to ∞,

γ) infinitely many solutions with either exactly one maximum or exactly one minimum, the range forming a half-line.

The ranges of solutions corresponding to types β) and γ) are illustrated in fig. 7.2a).

If f has three zeros $\xi_1 < \xi_2 < \xi_3$ (see the alternative b) above), as consequence of (7.6) the slopes at ξ_1 and ξ_3 are negative and $f'(\xi_2) > 0$. Therefore V has two maxima (at ξ_1 and ξ_3) separated by a minimum (at ξ_2), see fig. 7.2b). To each C obeying $V(\xi_2) < C \leq V(\xi_1)$, $C \leq V(\xi_3)$ there corresponds a stationary solution, its range forming an interval $R = [a,b]$ of finite, non-zero length. It can be shown that these solutions are either periodic in space, namely if $V'(a) \neq 0 \neq V'(b)$, or approach asymptotically the values, where V' vanishes.

<u>Lemma 7.2.</u> If, in the case of excitatory coupling (i.e. U a nonnegative, increasing, bounded function), the function f defined by (7.6) has exactly three zeros $\xi_1 < \xi_2 < \xi_3$ then, besides infinitely many unbounded solutions characterized by β) and γ) of lemma 7.1, there exist infinitely many stationary solutions of (7.7) which are periodic func-

tions of the space variable. Moreover in the structurally unstable case

$$\int_{\xi_2}^{\xi_1} f(\xi)\, d\xi = \int_{\xi_2}^{\xi_3} f(\xi)\, d\xi, \tag{7.10}$$

there is a monotone increasing bounded stationary solution.

Altogether only in the case b) (excitatory coupling) non-trivial bounded stationary solutions do exist. It is important to find out whether these solutions are stable with respect to small perturbations, because only then they have a chance to bo observed experimentally. The concept of stability used here is defined as follows.

Definition 7.3. For a function $u = u(s,t)$, defined for all $s \in \mathbb{R}$ and some t, let $\|u(.,t)\|_0 = \sup \{ u(s,t) : s \in \mathbb{R} \}$. A solution $v: \mathbb{R} \times [0,\infty) \to \mathbb{R}$ of (7.7) is called C^0-stable if, given $\varepsilon > 0$, there is a $\delta > 0$ such that for any initial condition $\tilde{v}(s,o)$ satisfying

$$\|\tilde{v}(.,o) - v(.,o)\|_0 < \delta$$

the corresponding solution $\tilde{v}(s,t)$ exists on $\mathbb{R} \times (0,\infty)$ and $\|\tilde{v}(.,t) - v(.,t)\|_0 < \varepsilon$ holds for all $t > 0$.
If, moreover,

$$\lim_{t \to \infty} \|\tilde{v}(.,t) - v(.,t)\|_0 = 0,$$

then v is called C^0-asymptotically stable.

These stability concepts are straightforward generalizations of those for ordinary differential equations. We state the following non-trivial result.

Lemma 7.4. (Fife, 1979). Every non-constant bounded stationary solution of (7.7) with a maximum or minimum at a finite value of s is not C^0-stable.

As consequence of the last three lemmas all non-constant bounded stationary solutions of (7.5) are unstable in the C^0 sense or structurally unstable. The stability of a constant solution $\bar{v}$ depends on the slope $f'(\bar{v})$. More generally it follows from the maximum principle that $\bar{v}$ is C^0-stable as a solution of (7.7) if and only if it is a stable solution of the ordinary differential equation

$$du/dt = f(u), \tag{7.11}$$

(see Fife, 1979, theorem 4.10).

Obviously $\bar{v}$ is a stable solution of (7.11) if $f'(\bar{v})<0$, and it is unstable if $f'(\bar{v})>0$. Therefore we have:

Lemma 7.5. If there is only one constant solution of (7.5) then this solution is C^0-stable. If there are exactly three constant solutions then the smallest and the largest ones are C^0-stable, and the intermediate one is C^0-unstable.

7.4. Thresholds in bistable tissues

The preceding section has shown that in the case of lateral excitation it may happen that there are two stable constant states. The question arises, given an initial distribution $\phi(s) = v(s,0)$, $-\infty<s<\infty$, of the potential $v(s,t)$, whether with increasing time the potential will approach one of the constant states and, if so, which of the two constants will be the limit.

It will turn out that the unstable stationary solutions can be used to describe regions of attraction of the stable constant states.

The following results are essentially due to Aronson and Weinberger, 1975, and Fife and McLeod, 1977.

For an easy representation assume that by rescaling the variables the three constant solutions have the values $0, \gamma$, and 1, $0<\gamma<1$. This means the function V, defined by (7.9) has two maxima at 0 and 1 and a minimum at γ.

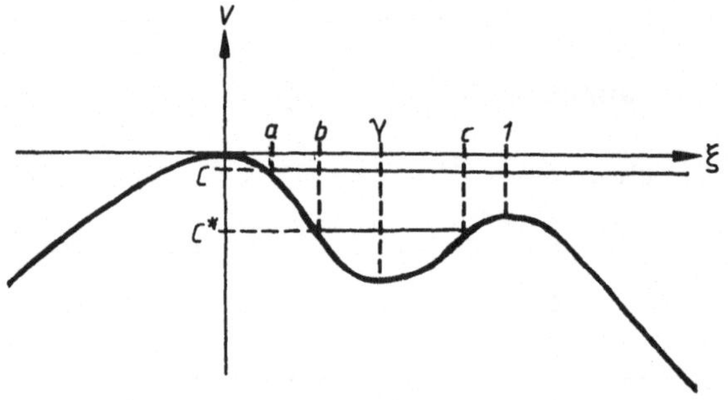

Fig. 7.3

Moreover let $V(0)>V(1)$, (the reverse case, $V(0)<V(1)$, is treated analogously). Fix some value C between V(0) and V(1), $V(0)>C>V(1)$. Then, according to the preceeding section, there is a stationary solution $v_0(s)$ with range $R = [a,\infty)$, where a is determined by $0<a<\gamma$ and $V(a) = C$ (note that there is a family of such solutions, since $v_0(\pm s-\varkappa)$ also solves (7.8) for each constant $\varkappa$). v_0 has exactly one minimum, say $s = s_0$, $v_0(s_0) = a$, and is monotone in the intervals $(-\infty,s_0]$, $[s_0,\infty)$. Let J denote the interval where $v_0(s) \leq 1$ holds.

Theorem 7.6. (Aronson, Weinberger, 1975). Let the bounded initial condition v(s,o) satisfy

$v(s,o) \leq v_o(s)$ for all $s \in J$.

Then the corresponding solution of (7.7) converges to the constant solution O:

$$\lim_{t \to \infty} v(s,t) = 0, \quad -\infty < s < \infty . \qquad (7.12)$$

This theorem can be interpreted as follows: With the condition $V(0) > V(1)$, a sufficiently large local depression of the potential (specified by the condition in the theorem) will lead to a global depression (i.e. the final state is O). Symmetrically, if $V(0) < V(1)$, a local excitation of sufficient strength will spread out to a global constant excitation (i.e. $\lim_{t \to \infty} v(s,t) = 1$ for all s).

A similar threshold condition is expressed in the following

Theorem 7.7. (Fife, McLeod, 1977): Assume $V(0) > V(1)$. Let η be a positive number. If the bounded initial condition $v(s,o)$ satisfies $v(s,o) \leq \gamma - \eta$ on an interval of sufficient length, then the corresponding solution $v(s,t)$ has the limit O, i.e. (7.12) holds.

Remark. The methods used to prove theorems 7.6 and 7.7 also show that in the case of lateral inhibition (more generally if f is monotonic decreasing) the unique constant solution of equation (7.7) is globally stable; more exactly, the solution $v(s,t)$ corresponding to any bounded initial condition $v(s,o)$, $-\infty < s < \infty$, converges to the constant solution as $t \to \infty$, uniformly in s.

A statement complementary to theorem 7.6 is available describing attractive domains of the stable constant solution 1 for the situation $V(0) < V(1)$.
However, the boundaries are now constructed by use of the periodic stationary solutions existing according to lemma 7.2: Fix some number C^* between $V(\gamma)$ and $V(1)$: $V(\gamma) < C^* < V(1)$, see fig. 7.3. Let $v_1(s)$, $-\infty < s < \infty$, denote any periodic stationary solution with range $[b,c]$, where b,c

are determined by $0<b<\alpha<c<1$ and $V(b) = C^* = V(c)$. Let I_1 be a closed interval containing just 4 zeros of v_1, two of them being endpoints of I_1, and the midpoint of I_1 being a minimum of v_1.

Theorem 7.8. (Aronson, Weinberger, 1975): Assume $V(0) > V(1)$. Then any solution $v(s,t)$ of (7.7), the initial condition of which is bounded and obeys

$$v(s,0) \gneq v_1(s) \text{ for } s\in I_1, \quad v(s,0) \gneq \alpha \quad \text{for } s\notin I_1,$$

converges to the constant solution 1 as $t \to \infty$:

$$\lim_{t\to\infty} v(s,t) = 1, \quad -\infty<s<\infty.$$

7.5. Traveling fronts

In the situation where f has three zeros, $\xi_1 < \xi_2 < \xi_3$, besides the two stable constant solutions of (7.7) $v(s,t) \equiv \xi_1$ and $v(s,t) \equiv \xi_3$, there is another type of stable solution : a traveling front.

A traveling front is a solution of (7.7) which can be written in the form

$$v(s,t) = \varphi(s-ct), \tag{7.13}$$

with a function $\varphi:\mathbb{R}\to\mathbb{R}$ having distinct constant asymptotes at infinity: the limits $\varphi(\infty) = \lim_{z\to\infty} \varphi(z)$ and $\varphi(-\infty) = \lim_{z\to-\infty} \varphi(z)$ exist and

$$\varphi(\infty) \neq \varphi(-\infty).$$

The two asymptotes necessarily coincide with two of the constant solutions.

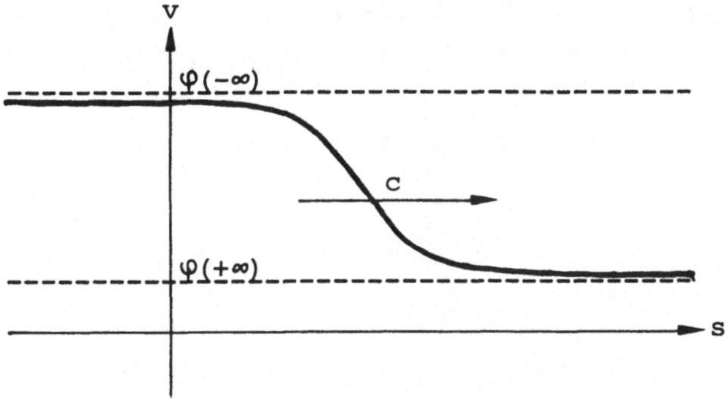

Fig.7.4.A traveling front

φ as a function of z describes the time-invariant shape of the front. The front travels with the velocity c to the right or to the left according as $c>0$ or $c<0$ respectively. As a solution of (7.7) φ has to solve the ordinary differential equation

$$\varphi'' + c\varphi' + f(\varphi) = 0,$$

where $\varphi(-\infty)$ and $\varphi(\infty)$ are two different zeros of f.

The existence problem for traveling fronts has been investigated in various contexts, e.g. in population genetics, in the theories of flame propagation and of propagation of impulses in nerve fibers. Detailed expositions are contained in (Hadeler-Rothe, 1975) and (Fife, 1979).

To summarize:

Theorem 7.9. In the case of lateral excitation let the differentiable function f, defined in (7.6), have exactly three zeros $\xi_1<\xi_2<\xi_3$. Let $i \neq j$ be two numbers from the set $\{1,2,3\}$.

(a) If $\{i,j\} = \{1,3\}$ then there exists a unique (up to translations in space) traveling wave solution (7.13) of (7.7) satisfying $\varphi(-\infty) = \xi_i$, $\varphi(\infty) = \xi_j$. Its velocity c obeys $c \gtreqless 0$ according as

$$|\int_{\xi_2}^{\xi_1} f(\xi)\, d\xi| \lesseqgtr |\int_{\xi_2}^{\xi_j} f(\xi)\, d\xi|. \qquad (7.14)$$

(b) If $2 \in \{i,j\}$ then there is a number $c^* > 0$ such that to each $\gamma \geq c^*$ there belongs a traveling wave (7.13) satisfying $\varphi(-\infty) = i$ and $\varphi(\infty) = j$. Its velocity is $c = \gamma$, if $j = 2$, and $c = -\gamma$, if $i = 2$.

The uniqueness in part a) of the theorem is restricted to the shape and velocity of the front, since with $v(s,t)$ also $v(s+s_o, t+t_o)$ is a solution of (7.7) for all constants s_o, t_o. As a rule the front travels into that direction which leads to an approach to the constant solution having the larger attractivity region. Analogous to the stationary solutions the problem arises which, if any, of the traveling fronts are stable.

It is heuristically clear that the fronts described in part b) of theorem 7.9 are unstable, since one of their asymptotes is the unstable constant solution $v(s,t) \equiv \xi_2$. Indeed the instability can be concluded from a very general result, obtained by Fife and McLeod, which also settles the stability problem for the fronts of type a) in theorem 7.9.

<u>Theorem 7.10</u>. (Fife and McLeod, 1977, McLeod and Fife, 1979). Let the hypothesis of theorem 7.9 hold. If the initial condition $v(s,0)$ satisfies

$$\limsup_{s \to -\infty} v(s,0) < \xi_2 < \liminf_{s \to \infty} v(s,0) \qquad (7.15)$$

then the solution $v(s,t)$, $t > 0$, of (7.7) obeys

$$|v(s,t) - \varphi(s-ct)| < C e^{-wt},$$

where φ is some traveling wave solution with $\varphi(-\infty) = \xi_1$, $\varphi(\infty) = \xi_3$ and C, w are constants.
Obviously a similar result holds if the roles of ξ_1 and ξ_3 are interchanged and the inequalities (7.15) are reversed.

Corollary 7.11. The traveling wave solutions in theorem 7.9(a) are C^0-stable, those in 7.9(b) are C^0-unstable.

Proof. If ψ is a traveling wave solution with $\psi(-\infty) = \xi_2$ and $\psi(\infty) = \xi_3$ (i.e. ψ is of type 7.9 (b)) then the initial condition $v(s,0) = \psi(s)-\varepsilon$ with a constant $0<\varepsilon<\xi_3-\xi_2$ obeys the inequalities (7.15). By theorem 7.10 the corresponding solution v(s,t) converges to a traveling front of type 7.9(b). Therefore ψ is unstable.

The stability result follows from the fact, implicit in the proof of theorem 7.10, that the constants C,w and the front φ depend continuously on the initial condition. Q.E.D.

It should be noted that the first stability results for traveling fronts in diffusion equations (7.7) with two stable constant solutions seem to have been obtained by Ya.I. Kanel' in the early sixties (for literature see Hadeler, 1976, Fife, 1979).

7.6. Diverging pairs of fronts (spread of excitation or depression)

Perhaps more important for applications than the phenomenon of traveling fronts is the possibility that a local excitation in a network may spread to the whole tissue, or the reverse process, that a totally excited network may pass into a depressed state by application of a local inhibition.

Consider again equation (7.5) for a one-dimensional tissue

with lateral excitation (U increasing) and a constant value I, such that the function f in (7.7) has exactly three zeros $\xi_1 < \xi_2 < \xi_3$ (see fig. 1). Then there is a constant state of depression, $v(s,t) \equiv \xi_1$, and a constant excited state of the network, $v(s,t) \equiv \xi_3$. Both states are stable.

Assume

$$\left| \int_{\xi_1}^{\xi_2} f(\xi)\, d\xi \right| < \int_{\xi_2}^{\xi_3} f(\xi)\, d\xi . \tag{7.16}$$

Moreover, assume the network to be at time t = 0 in the depressed state outside a bounded domain, i.e. $v(s,0) = \xi_1$ for sufficiently large $|s|$, and in a certain interval, say $L_1<s<L_2$, $v(s,0) > \xi_2+\varepsilon$ holds for some constant $\varepsilon>0$ (application of a brief local stimulus of sufficient strength can lead to such a situation). Then the solution $v(s,t)$, $t>0$, of (7.7) will approach a pair of diverging fronts as indicated in fig. 7.2 (at least if L_2-L_1 is large enough, depending on ε).

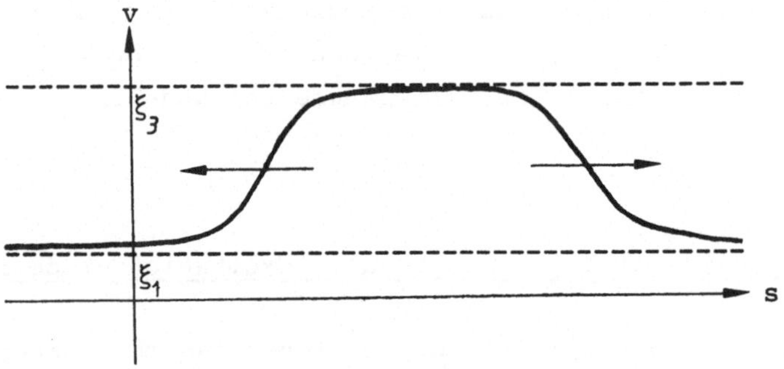

Fig. 7.5. Spread of excitation

The local excitation will spread onto the whole tissue as $t \to \infty$. This phenomenon is in line with the considerations in section 7.4. In particular the diverging front pair is a stable configuration (for mathematical details see Fife and McLeod, 1977).

When the inequality (7.16) is reversed a spread of depression becomes possible:

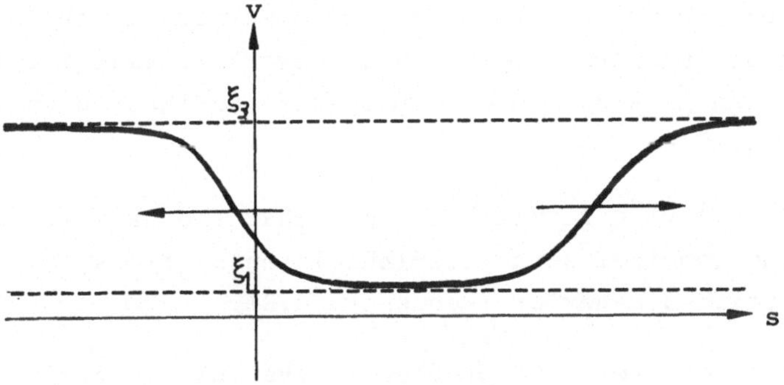

Fig. 7.6. Spread of depression

The roles of ξ_1 and ξ_3 are simply interchanged (fig. 7.6). However, if the inequality (7.16) holds and the net is in the excited state ξ_3 then a local depression cannot take the net into the depressed state ξ_1 (as follows from the results in section 7.4).

8. Homogeneous tissues with lateral excitation and self-inhibition

8.1. The model

In this chapter we discuss the possibility of a tissue in

which a local excitation can be propagated in such a way that

(i) the excitation moves through the whole tissue in an undamped manner, and

(ii) the excitation remains local, i.e. after the excitation wave has passed the tissue returns to its resting state of low activity.

The preceding sections suggest that a single layer of mutually coupled excitatory neurons, in particular the model (7.5), is unable to produce this type of behavior, if no further mechanism is assumed. Clearly a network realizing only lateral inhibition is also unable to propagate excitation in an undamped way.

In the following it will be shown that the lateral excitatory coupling combined with self-inhibition suffices to produce the described behavior (see sect. 8.3).

We shall not treat the problem in its full generality. Rather as a prototype, a specific and most simple model will be choosen allowing us to make use of the extensive mathematical studies in the context of the Hodgkin-Huxley model for the generation of nervous impulses in nerve axons.

Let us start with a homogeneous, one-dimensional tissue of cells with excitatory couplings.
The excitatory interaction is then modelled by the integral equation

$$v(s,t) = \int_{\mathbb{R}} U(v(s',.)) * h(|s-s'|,.)ds' + E, \qquad (8.1)$$

$-\infty<s<\infty, t>0,$

where again E denotes a (spatially and temporally) constant external input.
Self-inhibition is added to the process in the following way.

Assume the synaptic transmission is subject to exhaustion due to previous synaptic activity. Therefore instead of $U(v(s',t'))$ a quantity $U(v(s',t')) - w(s',t')$ becomes effective postsynaptically, where

$$w(s,t) = \int_{-\infty}^{t} v(s,t')\,\tilde{h}(t-t')dt'. \qquad (8.2)$$

Equation (8.1) is generalized to

$$v(s,t) = \int_{\mathbb{R}} [U(v(s',.)) - w(s',.)] * h(|s-s'|,.)ds'+E. \qquad (8.3)$$

Assuming h to be of the form (7.2) and $\tilde{h}(t) = b \exp(-dt)$ with constants b, d>0, the right hand sides of the integral equations (8.2,3) may be differentiated to obtain the system of differential equations

$$v_t - Dv_{ss} = f(v) - w, \qquad (8.4a)$$

$$w_t = bv - dw, \qquad (8.4b)$$

where

$$f(v) = U(v) - \alpha v + I, \quad I = \alpha E \qquad (8.4c)$$

System (8.4) is formally identical with the Fitzhugh-Nagumo equations (Fitzhugh, 1961, Nagumo et al. 1962) proposed as a simplification of the Hodgkin-Huxley model for nerve impulses. Note that Fisher's equation (7.7) is a special case ($b = 0 = d$, $w \equiv 0$) of the Fitzhugh-Nagumo system. In the space-independent situation ($v_{ss} \equiv 0$) the Fitzhugh-Nagumo system can be viewed as a specialization of Liénard's equation (for details see Hadeler et al., 1976 and the following section).

8.2. Spatially homogeneous behavior: bulk oscillations

In a first approach we look for the spatially homogeneous solutions of (8.4) obeying

$$dv/dt = U(v) - \alpha v - w + I, \tag{8.5a}$$

$$dw/dt = bv - dw. \tag{8.5b}$$

It is supposed that $U : \mathbb{R} \to \mathbb{R}$ is a monotonic increasing, continuously differentiable, and bounded function. The monotonicity condition reflects the excitatory coupling in the tissue. Moreover the nonrestrictive assumption

$$\lim_{|\xi| \to \infty} U'(\xi) = 0$$

is made for technical reasons.
System (8.5) can be transformed into a single second order equation

$$\ddot{v} + h(v)\,\dot{v} + g(v) = 0 \tag{8.6}$$

of Liénard type:

Differentiate (8.5a) with respect to t, then substitute $w = dw/dt$ by the right hand side of (8.5b); finally w is eliminated by (8.5a). It turns out that

$$h(v) = \alpha + d - U'(v), \tag{8.7}$$

$$g(v) = d\,((\alpha + b/d)v - U(v) - I). \tag{8.8}$$

Constant states $(\bar{v},\bar{w})$ are determined by $g(\bar{v}) = 0$, $\bar{w} = d\bar{v}/b$, therefore:

Lemma 8.1. System (8.4) has a unique constant solution if and only if the equation

$$(\alpha+b/d)\bar{v} - U(\bar{v}) = I$$

has a unique solution $\bar{v}$. In particular $\bar{v}$ is unique if $U'(\xi) < \alpha+b/d$ for all ξ.

When the first condition of the lemma fails, in general at least three constant solutions are present. In this situation all the phenomena of chapter 7 are likely to occur, however we shall not verify this conjecture. Throughout this chapter we shall only be concerned with the situation where there is a single constant state.

The stability of $(\bar{v},\bar{w})$ with respect to space-independent perturbations is governed by the roots of the characteristic equation

$$\lambda^2 + h(\bar{v})\lambda + g'(\bar{v}) = 0,$$

$$\lambda_{1,2} = -h(\bar{v})/z \pm \sqrt{h^2(\bar{v})/4 - g'(\bar{v})}.$$

It follows from the uniqueness of $\bar{v}$ that $g'(\bar{v}) > 0$. Therefore, $(\bar{v},\bar{w})$ as a solution of (8.5), is asymptotically stable if $h(\bar{v}) > 0$, and it is unstable if $h(\bar{v}) < 0$.

Lemma 8.2. Assume (8.4) has a unique constant solution $(\bar{v},\bar{w})$. Then $(\bar{v},\bar{w})$ is stable against spatially constant perturbations, if $\alpha+d > U'(\bar{v})$. However, if $U'(\bar{v}) > \alpha+d$, then $(\bar{v},\bar{w})$ is unstable.

It is well known that the Liénard equation (8.6) has non-constant periodic solutions if the following conditions are satisfied (for a proof see e.g. Knobloch and Kappel, 1974):

(i) $g,h:\mathbb{R}\longrightarrow\mathbb{R}$ are continuously differentiable functions,
(ii) there is a steady state $\bar{v}$ such that

$$g(\xi) > 0 \quad \text{for} \quad \xi > \bar{v}, \; g(\xi) < 0 \quad \text{for} \quad \xi < \bar{v},$$

(iii) $h(\bar{v}) < 0$ and $h(\xi) > 0$ for large $|\xi|$,

(iv) the integrals $G(x) = \int_0^x g(\xi)d\xi$ and $H(x) = \int_0^x h(\xi)d\xi$ obey

$$\lim_{|x|\to\infty} G(x) = \infty, \; \lim_{x\to\infty} H(x) > 0, \; \lim_{x\to-\infty} H(x) < 0.$$

In our context condition (ii) reduces just to the uniqueness of the steady state and condition (iii) corresponds to the instability condition in Lemma (8.2). The other conditions being trivially satisfied,we obtain

<u>Theorem 8.3</u>. System (8.4) has a non-constant spatially homogeneous periodic solution, if there is a unique constant solution $(\bar{v},\bar{w})$ obeying $U'(\bar{v}) > \alpha + d$.

It is an open problem whether the periodic solution of this theorem is uniquely determined. However as a consequence of the Poincaré-Bendixon theory for motions in the plane at least one of the nonconstant periodic solutions of system (8.5) is stable. Genericly such a stable periodic orbit is a limit cycle. Moreover every non-constant solution of (8.5) converges towards one of the non-constant periodic solutions.

One might look at these periodic solutions as candidates for <u>bulk oscillations</u> in the tissue, but then the conjecture has to be verified that they are also stable against spatially inhomogeneous perturbations, a problem which is unsolved.
Bulk oscillations,where the potentials of all neurons in the tissue oscillate at a common frequency, phase and amplitude seem

to be an important mode of neural activity. Cardiac tissue and respiratory neurons are only the most obvious examples. The system (8.4) under the conditions of theorem 8.3 can be used as a most simple paradigm how such behavior can be generated by interactive cell populations.

8.3. Traveling pulses

A traveling pulse is a non-constant solution $(v(s,t), w(s,t))$, $s \in \mathbb{R}$, $t \in \mathbb{R}$, of equation (8.4) which can be written in the form

$$\begin{pmatrix} v(s,t) \\ w(s,t) \end{pmatrix} = \begin{pmatrix} \varphi(s-ct) \\ \psi(s-ct) \end{pmatrix} = \Phi(z),$$

$z = s-ct$, and obeys the boundary conditions

$$\lim_{z \to -\infty} \Phi(z) = \lim_{z \to \infty} \Phi(z). \tag{8.9}$$

Remember that for a traveling front these limits have to assume different values. The function $\Phi(z)$, $-\infty<z<+\infty$, describes the shape of the pulse, and c is the traveling velocity of the pulse ($c>0$ rightward moving wave, $c<0$ leftward moving wave, $c = 0$ stationary wave).

Evidently pulse solutions of (8.4) as a model for nerve impulses have attracted much attention and we shall review some of the mathematical results of interest in the present context.

Since the pulse is asymptotically constant the limits in (8.9) correspond to a constant solution of (8.4), i.e.

$$\Phi(-\infty) = \Phi(\infty) = (\bar{v},\bar{w}). \tag{8.10}$$

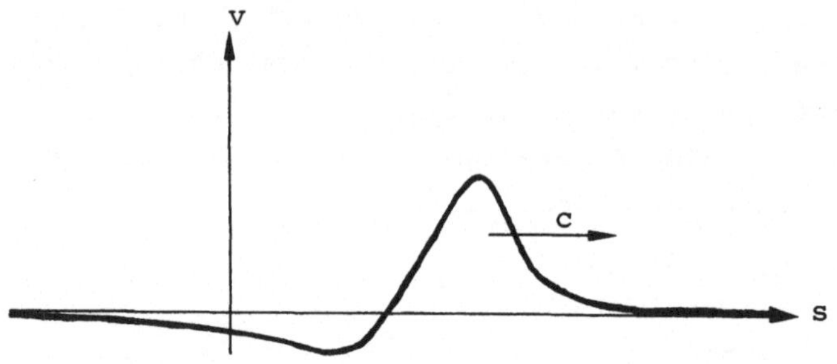

Fig. 8.1. A traveling pulse

The pulse solutions of (8.4) are characterized as those non-constant solutions $\Phi(z)$ of the ordinary system

$$D\varphi'' + c\varphi' + f(\varphi) = \psi \tag{8.11a}$$

$$-c\psi' + b\psi/d = b\varphi \tag{8.11b}$$

(with $\varphi' = d\varphi/dz$ etc.) which have the same limit as $z \to \pm\infty$. Such solutions are called <u>homoclinic orbits</u> of (8.11).

There are interesting existence results for rather general classes of functions f obtained by Carpenter, 1977, Conley, 1975 and Hastings, 1976. The essential assumptions are that f has three zeros $\xi_1 < \xi_2 < \xi_3$ such that $f'(\xi_1) < 0$, $f'(\xi_2) > 0$, $f'(\xi_3) < 0$ and

$$\left|\int_{\xi_1}^{\xi_2} f(\xi)d\xi\right| < \int_{\xi_2}^{\xi_3} f(\xi)\,d\xi, \tag{8.12}$$

see fig. 7.1.

Unfortunately they could prove their theorems only in restricted domains of parameter space, specificly d = 0 and b "sufficiently small".

Much more details and global results were derived by McKean, 1970, and Rinzel and Keller, 1973, who restricted their investigation to piecewise linear functions f of the form

$$f(\xi) = H(\xi - a) - \xi, \qquad 0 < a < \frac{1}{2}, \tag{8.13}$$

H = Heaviside function.

In other words they considered the special case U(v) = = H(v-a) of the parabolic system (8.4). Note that $\alpha = D = 1$ in (8.4) without loss of generality by rescaling the variables. The restriction for the parameter a in (8.13) is just choosen to satisfy the relation (8.12).

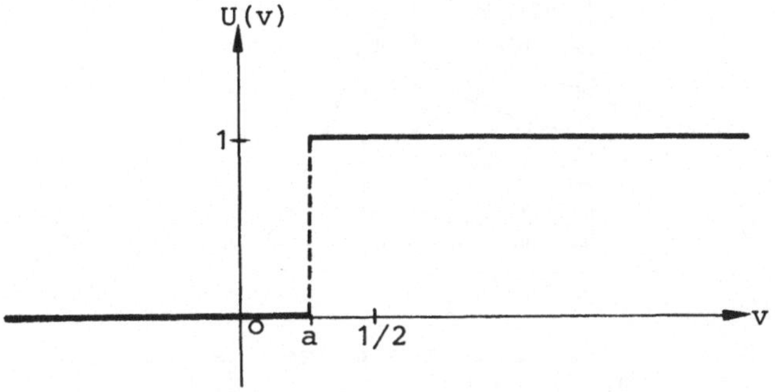

Fig. 8.2. U(v) = H(v-a)

The three authors additionally assume d = 0. Then they prove

(i) For each $b > 0$ there is a number $a_b < \frac{1}{2}$ such that for each pair (a,b) with $0 < a < a_b$ there are exactly two traveling pulse solutions of (8.4) with positive speeds,

(ii) the speeds of the two pulses are different. For any fixed b the speed c of the fast pulse is a decreasing function of the threshold a, and the slow pulse is an increasing function of a in the interval $0 < a < a_b$ (see fig. 8.3),

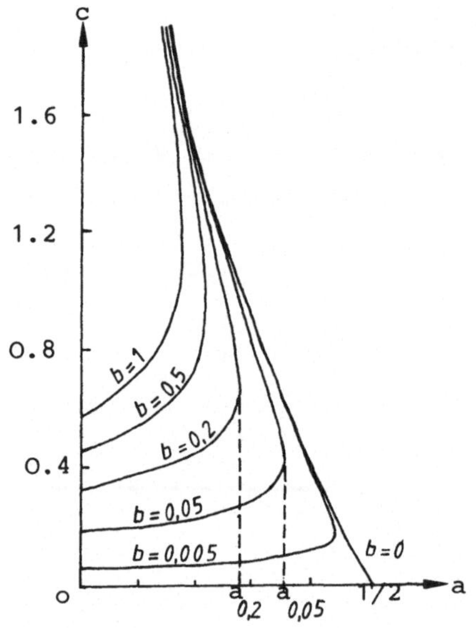

Fig. 8.3. Speeds c of the pulses as functions of a for fixed values b = 0,0.005.0.05,0.2,0.5, 1.0 (redrawn from Rinzel and Keller, 1973)

(iii) the pulse with the lower speed is unstable against small perturbations.

The numerical calculations (and also the analytical results of Carpenter, Conley and Hastings) so far obtained for other functions f are all consistent with these three propositions. The parameter a serves as a measure for the difference of the two integrals in (8.12).

There is strong indication (yet no proof even for f as in (8.13)) that the fast pulses are stable, and therefore are important from an experimental point of view.
Rinzel and Keller, 1973, also determined the amplitude (= sup $\varphi(z)$, $-\infty < z < \infty$) and width (= proportion of pulse above threshold) of the pulses as functions of a and b. Fig. 8.4 and fig. 8.5 portray their calculations.

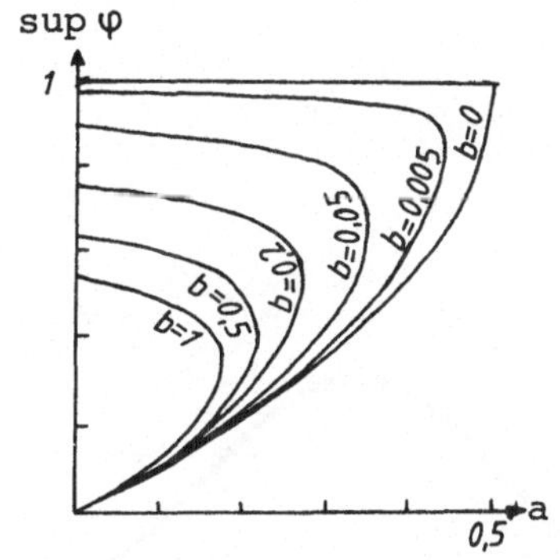

Fig. 8.4. Pulse heights as functions of a for various values of b (redrawn from Rinzel and Keller, 1973)

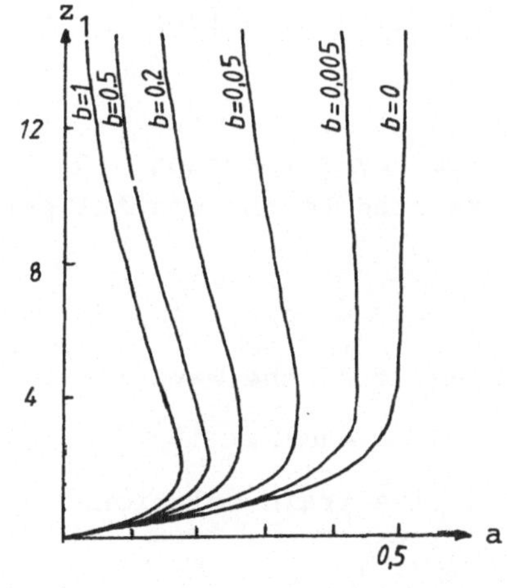

Fig. 8.5. Pulse widths as functions of a for various values of b (redrawn from Rinzel and Keller, 1973)

Note that for $b = 0$ the fast pulse degenerates to a traveling front, and the slow pulse to a standing pulse.

8.4. Traveling wave trains

Another type of traveling wave are wave trains. The shape of a wave train is a periodic function of the space variable. This pattern moves with a constant velocity c.

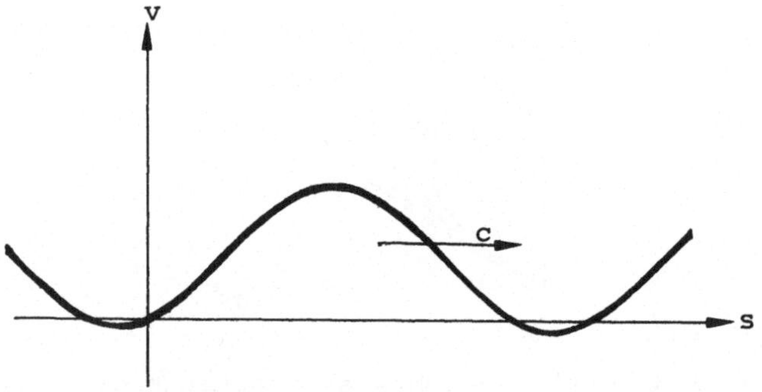

Fig. 8 .6. Traveling wave train

The wave is described by a function $\Phi(z) = \Phi(ks-\omega t)$. The wave number k is related to the spatial period σ (= wave length) by

$$\sigma = 2\pi/k,$$

the propagation speed c of the wave is

$$c = \omega/k.$$

With $z = ks - \omega t$ a wave train solution

$$\begin{pmatrix} v(s,t) \\ w(s,t) \end{pmatrix} = \begin{pmatrix} \varphi(z) \\ \psi(z) \end{pmatrix} = \Phi(z)$$

to the system (8.4) satisfies

$$k^2 D\varphi'' + \varphi' + f(\varphi) = \psi, \tag{8.14a}$$

$$\omega\psi - d\psi = -b\varphi, \tag{8.14b}$$

and

$$\Phi(z) = \Phi(z+2\pi). \tag{8.14c}$$

It is now very interesting that system (8.4) (lateral excitation combined with self-inhibition) has wave train solutions whenever there are traveling pulse solutions.

Again, existence of wave trains has been proved by Carpenter, 1977, Conley, 1975, and Hastings, 1976. A more complete picture has been obtained by Rinzel and Keller, 1973, for the piecewise linear case (8.13), where U equals Heaviside's step function:

Let a_b be defined as in proposition (i) of the previous section. Let $c_s(a,b)$ and $c_f(a,b)$ be the velocities of the slow and fast pulse, resp., existing for $b > 0$ and $0 < a < a_b$. Then, given any pair (a,b), $0 < a < a_b$, for each c obeying $c_s(a,b) < c < c_f(a,b)$ there exists a wave train solution $\Phi = \Phi_c$ (there may be other wave trains with c outside this interval). The wave lengths σ of these solutions fill a half-line $\sigma > \sigma_{min}(a,b)$. For these σ there are always two wave trains (with the same wavelength σ), but different speeds $c_s(a,b,\sigma) < c_f(a,b,\sigma)$. The limit relations

$$\lim_{\sigma\to\infty} c_s(a,b,\sigma) = c_s(a,b), \lim_{\sigma\to\infty} c_f(a,b,\sigma) = c_f(a,b)$$

show that the slow and fast pulses can be interpreted as

wave trains of infinite wave length.

The stability problem has not been completely settled. Numerical calculations suggest that there is a critical speed $c_o(a,b)$ such that all waves with velocity $c < c_o(a,b)$ are unstable. It may happen that both the slow and the fast wave train are unstable.
Rinzel and Keller derived a dispersion relation $\omega = \omega(k,a,b)$ describing for any given pair (a,b) of parameters the relation between the frequency ω and the wave number k for all members of the family of wave trains, which is parametrized by k.

Fig. 8 shows a numerical example of this relationship for $a = 0.3$, $b = 0.05$, $d = 0$. In general there are two frequencies associated with each wave number k, corresponding to the low and high speed traveling waves.

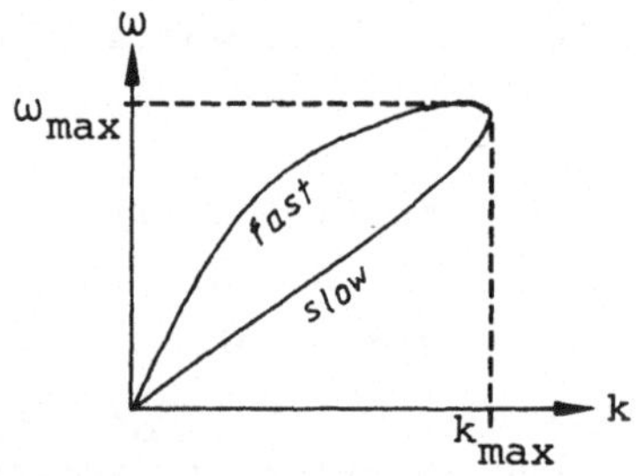

Fig. 8.7. Dispersion curve $\omega = \omega(k, a,b)$ for $a = 0.3$, $b = 0.05$, $d = 0$ of wave train solutions to (8.4) and (8.13) (redrawn from Rinzel, 1978)

9. Homogeneous networks with lateral inhibition and self-excitation: nonhomogeneous spatial patterns

9.1. The model

Here, in contrast with chapter 8, we assume that in a one-dimensional tissue the coupling is inhibitory, and that each cell has a domain of self-excitation.
The lateral spread of inhibition has again the temporal and spatial characteristics described by a function $h(s,t)$ as in (7.2). If in the expression for h the function U is simply linear, $U(a) = \beta a$, $\beta > 0$, then the total amount of inhibition exerted on the cells at location s at time t is

$$w(s,t) = \int_{\mathbb{R}} \beta v(s',.) * h(|s-s'|,.)ds'. \tag{9.1}$$

The potential $v(s,t)$ is assumed to obey the integral equation

$$v(s,t) = \int_{-\infty}^{t} [T(v(s,t')) - w(s,t')] \, e^{-\gamma(t-t')}dt' + E, \tag{9.2}$$

where $T:\mathbb{R} \rightarrow \mathbb{R}_+$ is a monotonic increasing, nonnegative, bounded function modeling self-excitatory effects and E is the value of a temporally and spatially constant external input. Clearly this model is a strong simplification of more realistic representations. E.g. the type of self-excitation can be looked upon as a limit case of local lateral excitation.

Differentiation of the equations (9.1), (9.2) results in the degenerate parabolic system

$$v_t = T(v) - \gamma v - w + I, \qquad (9.3a)$$

$$w_t = Dw_{ss} + \beta v - \alpha w, \qquad -\infty < s < \infty, \qquad (9.3b)$$

with constants $\alpha, \beta, \gamma > 0$, $I = \gamma E$.

This model is in a sense complementary to the model (8.4). Whereas (8.4) exhibits rich dynamic structures we shall see that (9.3) generates stable stationary patterns.

Systems similar to (9.3) have been considered by many authors in the theory of morphogenesis. This line of investigation was started by Turing, 1952. A great many of computer simulations and biological applications can be found in the work of Gierer and Meinhardt (see e.g. Meinhardt, 1977). The theory of dissipative structures also heavily relys on such models (see Nicolis and Prigogine, 1977). It is impossible to cite all references pertaining to this area (a review of some work is given by Fife, 1979). However, analytical methods and results are rather rare and restricted to simple questions or to simplified models.

An electronic circuit realization of the equations (9.3) has been proposed by Maginu, 1975, who also obtained some analytical results for a similar system.

In the morphogenetic context v and w denote the concentrations of chemical substances called morphogens. They are related to each other as "activator" (here v) and "inhibitor", and it is generally assumed that the inhibitor has a larger diffusivity than the activator as a prerequisite for the generation of stable spatial patterns. Here we have the extreme case that the diffusion constant for v equals 0.

The system (9.3) is one of the simplest imaginable for this purpose, and it is only chosen to enable extensive analysis and to clarify the main points.

For neural networks these models contribute to solve the problem how a homogeneous population of cells is able to maintain an inhomogeneous state of activity even if there is a spatio-temporally constant input (which may be zero).

9.2. Inhomogeneous stationary solutions

The stationary solutions of (9.3) in the interval $-\infty<s<\infty$ coincide with the solutions of the degenerate differential equation system

$$y = f(x), \tag{9.4a}$$

$$y'' - \alpha y = -x, \tag{9.4b}$$

where D has been normalized to 1 by rescaling, $y' = dy/ds$, $x = \beta(v-\bar{v})$, $y = w-\bar{w}$, $(\bar{v},\bar{w})$ the constant solution of (9.3) (assumed to exist uniquely).

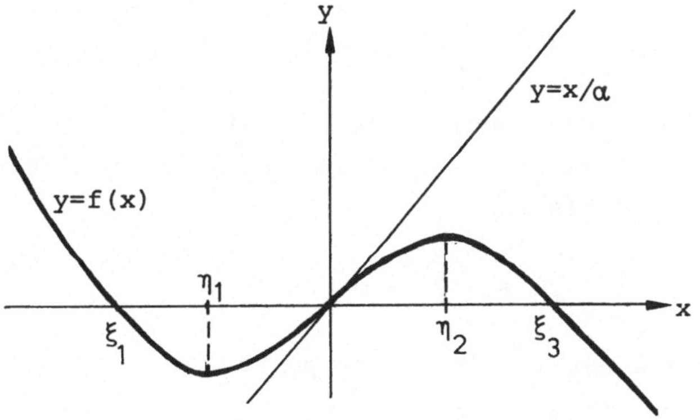

Fig. 9.1. A typical function f in (9.4)

Consider the situation where the differentiable function f has exactly three zeros $\xi_1 < \xi_2 = 0 < \xi_3$, is strictly monotone in each of the intervals $(-\infty,\eta_1)$, (η_1,η_2), (η_2,∞), and $\lim_{\xi\to\pm\infty} f(\xi) = \mp\infty$ resp. as depicted in fig. 9.1.
Let f be convex in the interval $(\xi_1,0)$ and concave in $(0,\xi_3)$.

Then (0,0) is the only constant solution of (9.4) if and only if

$$\alpha f'(0) < 1. \tag{9.5}$$

In order to determine all solutions of (9.4) three cases are distinguished:

<u>Case 1.</u> Let $(x(s), y(s))$, $-\infty<s<\infty$ be a solution of (9.4) such that $x(s_0) < \eta_1$ for some fixed $s = s_0$.
Let f_1 be the function f restricted to the domain $(-\infty,\eta_1]$. Since the solutions of (9.4) have to be continuous, there is an interval I around s_0, $s_0 \in I$, such that for all $s \in I$ the relation

$$x(s) = f_1^{-1}(y(s)) \tag{9.6}$$

holds. The boundary points of I may be $\pm\infty$ or a finite value s_1 with $x(s_1) = \eta_1$. Because of (9.6) and (9.4) the y-component satisfies

$$y'' + g(y) = 0, \tag{9.7}$$

where $g(y) = -\alpha y + f_1^{-1}(y)$, $y \geq f(\eta_1)$.
Integration of (9.7) results in

$$\frac{1}{2}(y')^2 + V(y) = E \tag{9.8}$$

for some constant E and $V(y) = \int_0^y g(\xi)\, d\xi$.
Equation (9.8), i.e.

$$y' = \pm \sqrt{2(E-V(y))}\ , \tag{9.9}$$

holds only as long as $y \geq f(\eta_1)$.

V is a strictly decreasing function of y. If $E > V_{max} = V(f(\eta_1))$, then, according to (9.9), $|y'| > +\sqrt{2(E-V_{max})} > 0$, therefore s_1 exists with $y(s_1) = f(\eta_1)$, $|y'(s_1)| > 0$, $x(s_1) = \eta_1$. This implies, because of (9.4a), a discontinuity of x at s_1. Therefore only constants $E \leq V_{max}$ lead to a solution. For $E < V_{max}$, $y' = 0$ exactly if $V(y) = E$. In this situation, because of (9.9) and the continuity of x, always $y(s) > f(\eta_1)$, and the solution component y(s) is monotone on the intervals $-\infty < s \leq s_2$, $s_2 \leq s < \infty$ for some number s_2, $\lim_{s\to\pm\infty} y(s) = \infty$. Therefore for each $E < V_{max}$ there is an unbounded stationary solution. For $E = V_{max}$ it is possible that x leaves the interval I at a finite value $s = s_1$, but nevertheless, because of (9.9), either $\lim_{s\to-\infty} x(s)$ or $\lim_{s\to\infty} x(s)$ is infinite, and hence the solution is also unbounded.

Case 2. Let $(x(s), y(s))$, $-\infty < s < \infty$, be a solution of (9.4) such that $\eta_1 < x(s_0) < \eta_2$ for some $s = s_0$.

Let f_2^{-1} denote the inversion of f in the domain $\eta_1 \leq s \leq \eta_2$ and $V(y) = \int_0^y (-\alpha\xi + f_2^{-1}(\xi))d\xi$, $f(\eta_1) \leq y \leq f(\eta_2)$. Then again the relations (9.8), (9.9) and $x(s) = f_2^{-1}(y(s))$ hold in an interval I around s_0 for some constant E. The situation is depicted in fig. 9.2.

Without loss of generality assume $V_{max} = V(f(\eta_2)) \geq V(f(\eta_1))$. V is monotone in the negative and positive parts of its domain and has a minimum at y = 0.
If $E > V(f(\eta_1))$ then because of (9.9) for some $s = s_1$ the function y(s) transcends the value $f(\eta_1)$ from above. Since

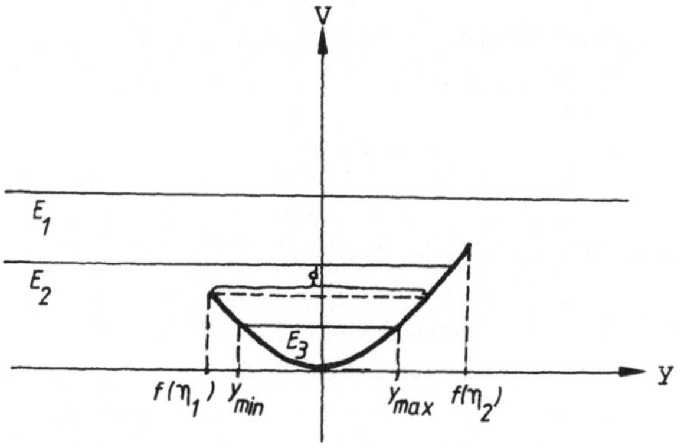

Fig. 9.2

$y'(s_1) < 0$ this again implies a discontinuity of the variable x. Therefore $E \lesssim V(f(\eta_1))$ is necessary.
The case $E = 0$ corresponds to the constant solution (0,0).
Finally let $0 < E \lesssim V(f(\eta_1))$. In this situation $(x(s), y(s))$ is a periodic solution. The range of y is the interval $[y_{min}, y_{max}]$, where y_{min} and y_{max} solve $V(\xi) = E$. As $E \to 0$ the amplitude converges to zero. The maximal amplitude d is associated with $E = V(f(\eta_1))$.

Case 3. Let $(x(s), y(s))$, $-\infty < s < \infty$, be a solution of (9.4) such that $\eta_2 < x(s_0)$ for some $s = s_0$. The discussion of this case is analogous to that of case 1.
In summary the following result has been obtained.

Theorem 9.1. Let the function f and the parameter α satisfy the conditions stated in the beginning of this section. Then the set of all bounded stationary solutions of (9.3) consists of the constant solution and a family of non-constant spatially periodic solutions. This family can be parametrized by the amplitudes of the periodic solutions. The amplitudes fill a bounded interval $(0,d]$, $d>0$ as determined above.

9.3. Stability

At first it will be shown how the stability of the constant solution of (9.3) depends on the various parameters. In particular it will turn out that the constant state can be unstable even if it is stable as a solution of the associated spatially homogenous system

$$\dot{v} = T(v) - \gamma v - w + I \tag{9.10a}$$

$$\dot{w} = \beta v - \alpha w, \tag{9.10b}$$

$\dot{v} = dv/dt$, $\dot{w} = dw/dt$ (compare with equation 8.5).
Assume T to be a differentiable, monotone increasing, and bounded function, satisfying

$$T'(\xi) < (\gamma + \beta/\alpha) \text{ for all } \xi \in \mathbb{R} \tag{9.11}$$

The last inequality guarantees that there is a unique constant solution $(\bar{v},\bar{w})$.
The linearization of (9.3) about $(\bar{v},\bar{w})$ is

$$u_t = (T'(\bar{v})-\gamma)\, u - z \tag{9.12a}$$

$$z_t = Dz_{ss} + \beta u - \alpha z. \tag{9.12b}$$

$(\bar{v},\bar{w})$ is asymptotically stable as a solution of (9.10), if

$$T'(\bar{v}) - \gamma - \alpha < 0 \tag{9.13}$$

and

$$\beta - \alpha(T'(\bar{v})-\gamma) > 0. \tag{9.14}$$

However, (9.14) follows from (9.11). In the following we assume (9.13) to hold.

Let us determine the solutions to (9.12) of the form

$$u = u_o \exp(ikx + \lambda t), \tag{9.14a}$$

$$z = z_o \exp(ikx + \lambda t) \tag{9.14b}$$

for arbitrary wave numbers $k \in \mathbb{R}$. (9.14) is a solution if and only if $\lambda = \lambda(k)$ is an eigenvalue and (u_o, z_o) an eigenvector of the matrix

$$J(k) = \begin{pmatrix} (T'(\bar{v}) - \gamma) & -1 \\ \beta & (-k^2 D - \alpha) \end{pmatrix}.$$

The solution $(\bar{v}, \bar{w})$ to (9.3) is unstable if $\text{Re}\,\lambda(k) > 0$ for at least one k.

Because of (9.13), trace $J(k) < 0$ for all k. Therefore $\text{Re}\,\lambda(k) > 0$ occurs if and only if $\det J(k) < 0$, i.e.

$$\beta - (k^2 D + \alpha)(T'(\bar{v}) - \gamma) < 0 \text{ for some } k, \tag{9.15}$$

(note that in this case $\lambda(k)$ is real).

Condition (9.15) holds if and only if

$$T'(\bar{v}) - \gamma > 0. \tag{9.16}$$

The instability generating waves (9.14) are those with wave numbers k obeying

$$k^2 > \frac{\beta - \alpha(T'(\bar{v}) - \gamma)}{D(T'(v) - \gamma)}.$$

To summarize:

<u>Theorem 9.2</u>. The constant solution $(\bar{v},\bar{w})$ of system (9.3) is unstable if the inequalities (9.13), (9.14) and (9.15) are satisfied.

It was this kind of analysis which led A.M.Turing to the conjecture that certain types of reaction-diffusion systems give rise to stable spatial patterns. All the many numerical experiments (for review see Nicolis and Prigogine, 1977) in the meantime performed have supported this view. Bifurcation techniques have been developed to prove the stability of small amplitude stationary solutions (see Crandall and Rabinowitz, 1973). As an example for applications to neural networks, see Ermentrout and Cowan, 1979. However, there has been no success (with except of a class of constructive examples , Fife, 1977), to prove rigorously the existence of stable large amplitude stationary patterns. Therefore it is an open problem, which of the periodic solutions in theorem 9.1 are stable and which are unstable. Work of Maginu, 1975, Rothe - de Mottoni, 1979, and Rothe, 1979, on an equation similar to (9.3) (however with bounded spatial domains) makes it quite unlikely that the solutions of (9.3) asymptotically depend on time. Therefore at least in view of theorem 9.2 one of the spatially periodic, but stationary, solutions should be stable.

9.4 Miscellaneous topics

The analysis of neural tissues in the last few chapters (6-8) has been very restricted. Most basic has been the restriction to simple layer tissues (m=1 in (1.19)) and the restriction to kernels of type (7.2), allowing the transition from integral equations to differential equations. We will indicate here briefly and without any claim for completeness some additions and extensions of the theory as it can be found in the literature. Double layered tissues have been studied by Wilson and Cowan, 1973, Cowan and Ermentrout 1978, Ermentrout and Cowan 1979, Tokura and Morishita, 1977. These authors also consider more general types of kernels. However, their results are generally based on numerical methods (computer simulation) or, if proved analytically, on techniques from bifurcation theory (mostly limited to bounded spatial domains). It is not a diminution of bifurcation analysis to remark that it can exhibit only "small amplitude phenomena" and that it covers only local regions in parameter space. Much and difficult work is necessary to generalize the global results on large amplitude solutions in chapters 7 - 9 to networks of a more general type. As a simplification of a 2-layer model a 1-layer model with an interaction kernel assuming positive and negative sign is discussed by Kishimoto and Amari, 1979.
Another restriction in the previous chapters has been the consideration of only one-dimensional spatial domains of the tissues. Whereas all phenomena occuring in lower dimensions also occur in higher dimensions, there are of course many types of behavior requiring at least a two-dimensional extension of the tissue. As an example we refer here to the investigations of spiral waves in the context of the Zhabotinsky-reaction by Winfree, 1978, which can also be interpreted in the light of neural network theory. Rotating waves are also studied by Greenberg and Hastings, 1978, in a discrete model with cells distributed over the grid points in the plane. Their equations, and those of Winfree, too, are

closely related to the Fitzhugh-Nagumo system discussed in chapt.8.

In this connexion the early paper of Wiener and Rosenblueth,1946, should be mentioned.They used a discrete diffusion model to explain spatial activities in cardiac muscles.

Another discrete model of several layers of neurons was used by Marr and Poggio, 1976, to simulate depth perception in binocular vision.

Their work is already directed towards applications to real spatially extended neural tissues. Still more closely related to experimental data are the investigations of Freeman, 1975, who obtained simultaneous recordings from cell arrays in the olfactory bulb. A reaction-diffusion model for cortical depression has been developed by Tuckwell and Miura,1978.

REFERENCES

Adam,A., 1968: Simulation of rhythmic nervous activities.II. Mathematical models for the function of networks with cyclic inhibition. Kybernetik (now: Biol. Cyb.) 5, 103-109

an der Heiden,U., 1976: Existence of periodic solutions of a nerve equation. Biol.Cybernetics 21, 37-39

an der Heiden,U., 1976: Stability properties of neural and cellular control systems. Math.Biosciences 31, 275-283

an der Heiden,U., 1978: Structures of excitation and inhibition. Lecture Notes in Biomathematics 21, 75-88 Berlin, New York :Springer Verlag

an der Heiden,U., 1979: Periodic solutions of a nonlinear second order differential equation with delays. J.Math.An.Appl.70, 599-609

Andersen,P., Gillow,M., Rudjord,T., 1966: Rhythmic activity in a simulated neuronal network. J.Physiol. 185, 418-428

Aplevich,J.D., 1967: Models of certain nonlinear systems. In: Caianiello,E.R.(ed.): Neural networks. Springer Verlag

Arbib,M.A., 1965: Brains, Machines and Mathematics. McGraw-Hill, New York

Aronson,D.G., Weinberger,H.F., 1975: Nonlinear diffusion in population genetics, combustion and nerve propagation. In: Proc. Tulane Progr. in Part. Diff. Equ., Lecture Notes in Math. 446, Springer, Berlin, 5-49

Barrett,J.N., Crill,W.E., 1974: Specific membrane properties of cat motoneurones. J. Physiol. (Lond.) 239, 301-324

Barrett,J.N., Crill,W.E., 1974: Influence of dentritic location and membrane properties on the effectiveness of synapses on cat motoneurones. J. Physiol. (Lond.) 239, 325-345

Bekesy,G.v., 1960: Neural inhibitory units of the eye and the skin. J.Opt. Soc.Am. 50, 1060-1070

Benevento,L.A., Creutzfeldt, O.D., Kuhnt, U., 1972: Significance of intra-cortical inhibitions in the visual cortex. Nature 238, 124-126

Beurle, R.L., 1956: Properties of a mass of cells which can regenerate impulses. Phil. Trans. Roy.Soc. (Lond.) B 240; 55,669

Blakemore,C., Carpenter,R., Georgeson,M., 1971: Lateral thinking about lateral inhibition. Nature 234, 418-419

Braess, D., 1967: Approximation mit Exponentialsummen. Computing 2, 309-321

Braess, D., 1970: Die Konstruktion der Tschebyscheff-Approximierenden bei der Anpassung mit Exponentialsummen. J.Approx.Theory 3, 261-273

Braess,D., 1973a, 1974: Chebyshev approximation by γ -polynomials. J.Approx. Theory 9, 20-43 and part II, same journal 11, 16-37

Braess,D., 1973b: Global analysis and Chebyshev approximation by exponentials. In "Approximation Theory" (G.G.Lorentz ed.), 277-282, Academic Press, New York

Browder,F.E., 1975: Singular nonlinear integral equations of Hammerstein type. In: Partial diff. equations and related topics, Proc. Tulane 1974, Lecture Notes in Math. 446, Springer Verlag

Butz,E.G., Cowan,J.D., 1974: Transient potentials in dendritic systems of arbitrary geometry. Biophys. J. 14, 661-689

Caianiello,E.R., 1961: Outline of a theory of thought processes and thinking machines. J. Theor. Biol. 1, 204

Caianiello,E.R., de Luca, A., Ricciardi,L.M., 1967: Reverberations and control of neural networks. Kybernetik 4, 10-18

Caianiello,E.R., Grimson,W.E.L., 1976, Methods of analysis of neural nets. Biol. Cyb. 22,1-6

Calvin,W.H., 1975: Generation of spike trains in CNS neurons. Brain Res. 84, 1-22

Carpenter,G.A., 1977: A geometric approach to singular perturbation problems with applications to nerve impulse equations. J.Diff. Equ. 23, 335-367

Carpenter,G.A., 1977: Periodic solutions of nerve impulse equations. J.Math. Anal. Appl. 58, 152-173

Carpenter,G.A., 1979: Bursting phenomena in excitable media. SIAM J.Appl. Math. 36, 334-372

Chen,C.F., von Baumgarten,R., Takeda,R., 1971: Pacemaker properties of completely isolated neurons in Aplysia californica. Nature (Lond.) New Biol. 233, 27-29

Cohen,H., 1976: Mathematical developments in Hodgkin-Huxley theory and its approximations.In: Lectures on Mathematics in the Life Sciences, Volume 8, 89-124

Coleman,B.D.,1971: A mathematical theory of lateral sensory inhibition. Arch. Rat. Mech. Anal. 43, 79-100

Coleman,B.D., Renninger,G.H.,1976: Periodic solutions of certain nonlinear integral equations with a time - lag. SIAM J. Appl. Math. 31, 111-120

Coleman,B.D.,Renninger,G.H. 1974: On the integral equations of the linear theory of recurrent lateral interaction in vision. Math. Biosc. 20, 155-170

Coleman,B.D., Renninger,G.H., 1975: Consequences of delayed lateral inhibition in the retina of Limulus I, II; J. theor. Biol. 51, 243-265

Coleman,B.D., Renninger, G.H., 1976: Theory of the response of the *Limulus* retina to periodic excitation. J. Math. Biol. 3, 103-119

Conley,C., 1975: On traveling wave solutions of nonlinear diffusion equations. Lect. Notes in Physics 38, Springer-Verlag, Berlin New York

Corduneanu,C., 1973: Integral equations and stability of feedback systems. Academic Press, N.Y.

Cowan,J.D., 1967: A mathematical theory of central nervous activity . Thesis, Univ. of London

Cowan,J.D., 1970: A statistical mechanis of nervous activity. In: Gerstenhaber,M. (ed.): Lecture on mathematics in the life sciences. Providence, R.I.; Amer. Math. Soc.

Cowan,J.D., Ermentrout,G.B., 1978: Some aspects of the "eigenbehavior" of neural nets. In: Studies in math. biology, (S.A.Levin ed.) pp. 67-117, Math. Association of America

Crandall,M.G., Rabinowitz,H.G., 1973: Bifurcation, perturbation of simple eigenvalues, and linearized stability. Arch. Rat. Mach. Anal. 52, 161-180

Cushing,J.M., 1975: An operator equation and bounded solutions of integrodifferential systems. SIAM J. Math. Anal. 6, 433-445

Cushing,J.M., 1977: Integrodifferential equations and delay models in population dynamics. Lecture Notes in Biomath. 20, Springer-Verlag, New York - Heidelberg

Cragg,B.G., Temperley,H.N.V.,1955
Brain 78, 304

Eccles,J.C., 1973: The understanding of the brain, McGraw Hill, New York

Elul,R., 1972: The genesis of the EEG. Int.Rev.Neurobiol. 15, 227-272

Ermentrout,G.B., Cowan, J.D., 1978: Large scale activity in neural nets (preprint)

Ermentrout,G.B., Cowan,J.D., 1979: Temporal oscillations in neural nets. J.Math.Biol.7, 265-280

Fargue,M.D., 1973: Reducibilité des systèmes héréditaires à des systèmes dynamiques. C.R. Acad. Sci. Paris B 277, 471-473

Fentress,J.C.(ed.),1976: Simpler networks and behavior, Sinauer Ass., Inc. Publ., Sunderland, Mass.

Feldman,J.L., Cowan,J.D., 1975: Large scale activity in neural nets. I: Biol. Cybernetics 17, 29-38
II: Biol. Cybernetics 17, 39-51

Feldman,J.L., 1976: A network model for control of inspiratory cutoff by the pneumotaxic center with supportive experimental data in cats. Biol. Cyb. 21, 131-138

Fife,P.C., 1977: Stationary patterns of reaction-diffusion equations. In: Nonlinear diffusion. Res. Notes in Math. 14, 81-121, Pitman, London

Fife,P.C., McLeod, J.B., 1977: The approach of solutions of nonlinear diffusion equations to traveling front solutions. Arch. Rat. Mech. Anal. 65, 335-361

Fife,P.C., 1979: Mathematical aspects of reacting and diffusing systems. Lect.Notes in Biomath.,Vol.28, Springer-Verlag

Fischer,B., 1973: A neuron field theory: mathematical approaches to the problem of large numbers of interacting nerve cells. Bull. Math. Biol. 35, 345-357

Fischer,B., Krüger,J., 1976: Mathematical principles in afferent visual neurons: differentiation, integration, and transient proportionality related to receptive fields and shift-effect. Bull. Math. Biol. 38, 253-267

Fitzhugh,R., 1961: Impulses and physiological states in theoretical models of nerve membrane. Biophys. J. 1, 445-466

Fitzgibbon,W.E., 1976: Nonlinear evolution operators and delay equations. In: Ord. and part.diff.equs. Dundee (Everitt and Sleeman eds.), Lect.Notes in Math.,Vol.564, Springer-Verlag

Freeman,W.J., 1975: Mass action in the nervous systems. Academic Press, New York-London

Friesen,W.O., Stent,G.S., 1977: Generation of a locomotory rhythm by a neural network with recurrent cyclic inhibition. Biol. Cyb. 28, 27-40

Fuortes, M.G.F.,Hodgkin,A.L., 1964: Changes in timescale and sensitivity in the ommatidia of Limulus. J. Physiol. 172, 239-263

Gainer,H., 1972: Electrophysiological behavior of an endogeneously active neurosecretory cell. Brain Res. 39, 403-418

Gierer,E., Meinhardt,H., 1972: A theory of biological pattern formation. Kybernetik 12, 30-39

Giesler,C.D., Goldberg, J.M., 1966: A stochastic model for the repetitive activity of neurons. Biophys. J. 6, 63-69

Goodwin,B.C., 1965: In: Weber, G.(Ed.): Advances in enzyme regulation. Vol. 3, p. 425. Oxford: Pergamon Press

Greenberg,J.M., Hastings, S.P., 1978: Spatial patterns for discrete models of diffusion in excitable media. SIAM J. Appl. Math. 34, 515-523

Griffith,J.S., 1963: A field theory of neural nets.I. Bull. Math. Biophys. 25, 111-120

Griffith,J.S., 1965: A theory of neural nets II. Bull. Math. Biophys. 27, 187-195

Griffith,J.S., 1968: Mathematics of cellular control processes I., II. J. Theor. Biol. 20, 202-216

Griffith,J.S., 1971: Mathematical neurobiology, Acad. Press NY

Hadeler,K.P., 1974: On the theory of lateral inhibition, Kybernetik 14, 161-165

Hadeler,K.P., Rothe,F., 1975: Traveling fronts in nonlinear diffusion equations. J. Math. Biol. 2, 251-263

Hadeler,K.P., an der Heiden,U., Schumacher,K., 1976: Generation of the nervous impulse and periodic oscillations. Biol. Cybernetics 23, 211-218

Hadeler,K.P., 1976: Nonlinear diffusion equations in biology. pp.163-206 in: Lect.Notes in Math.,Vol.564, Springer-Verlag

Hadeler,K.P., Tomiuk,J., 1977: Periodic solutions of difference-differential equations. Arch. Rat. Mech. Anal. 65, 87-95

Hadeler,K.P., 1979: Periodic solutions of the equation $x'(t) = -f(x(t), x(t-1))$. Math.Methods in the Appl. Sciences 1, 52-59

Hale,J., 1977: Theory of functional differential equations. Springer Verlag, New York - Heidelberg

Hastings,S.P., 1976: On traveling wave solutions of the Hodgkin-Huxley equations. Arch. Rat. Mech. Anal. 60, 229-257

Hastings,S.P., 1976: On the existence of homoclinic and periodic orbits for the Fitzhugh-Nagumo equations. Quart. J. Math. Oxford (2), 27, 123-134

Hastings,S., Tyson,J.J., Webster,D., 1977: Existence of periodic solutions for negative feedback cellular control systems. J. Diff. Equations 25, 39-64

Harmon,L.D., 1961: Neuromimes: Action of a reciprocally inhibitory pair. Science 146, 1323-1325 (1961)

Holden,A.V., 1976: Models of the stochastic activity of neurones. Springer-Verlag

Jenik,F., 1962: Electronic neuron models as an aid to neurophysiological research. Ergebn. Biol. 25, 206-245

Kammler,D.W., 1973: Existence of best approximates by sums of exponentials. J. Approx. Theory 9, 173-191

Kaplan,J.L., Yorke,J.A., 1977: On the nonlinear differential delay equation x'(t) = - f(x(t),x(t-1)). J. Differential Equs. 23, 293-314

Katchalsky,A., 1971: Biological flow structures and their relation to chemico-diffusional coupling. Neurosciences Research Program Bulletin 9, 397-413

Katchalsky,A., Rowland,V., Blumenthal,R. (eds.), 1974: Dynamic patterns of brain cell assemblies. Neurosciences Research Program Bulletin 12, 3-187

Katz,B., Miledi,R., 1967: A study of synaptic transmission in the absence of nerve impulses.J.Physiol. 192, 407-436

Kishimoto,K., Amari,S., 1979: Existence and stability of local excitations in homogeneous neural fields. J. Math. Biol. 7, 303-318

Kling,V., Szekely,G., 1968: Simulation of rhythmic nervous activities. I. Function of networks with cyclic inhibitions. Kybernetik 5, 89-103

Knight,B.W., 1973: Some questions concerning the encoding dynamics of neuron populations. Proc. 4th Internat. Biophysics Congress, Puschino, USSR, p.422-434

Knobloch, H.W. Kappel,F., 1974: Gewöhnliche Differentialgleichungen. Teubner Verlag, Stuttgart

Korn,A., von Seelen,W., 197?: Dynamische Eigenschaften von Nervennetzen im visuellen System. Kybernetik 2, 64-77

Kurokawa,T., Tamura,H., 1974: Networks of neural nuclei. Kybernetik 16, 69-77

Leibovic,K.N., 1972: Nervous System Theory. Academic Press, New York and London

Lopes de Silva,F.H., Hocks,A., Smits,H., Zetterberg,L.H.,1974: Model of brain rhythmic activity, the alpha-rhythm of the thalamus. Kybernetik 15, 27-37

Lovegrave,W., 1976: Inhibition in simultaneous and successive contour interaction in human vision. Vision Res. 16, 1519-1521

MacDonald,N., 1978: Time lags in biological models. Lecture Notes in Biomath. 27, Springer Verlag, New York-Heidelberg

MacGregor,R.J., Lewis,E.R., 1977: Neural modeling. Plenum Press, New York

Machemer,H., Eckert,R., 1973: Electrophysiological control of reversed ciliary beating in Paramecium. J.Gen.Physiol. 61, 572-587

Maffei,L., Fiorentini,A., Bisti,A., 1973: Neural correlate of perceptual adaption to gratings. Science 182, 1036-1038

Maginu,K., 1975: Reaction-diffusion equations describing morphogenesis. Math. Biosciences 27, 17-98

Marko,H., 1969: Die Systemtheorie der homogenen Schichten. Kybernetik 5, 221-240

Marr,D., Poggio,T., 1976: From understanding computation to understanding neural circuitry. MIT Art. Intell.Lab.

McCulloch,W.S., Pitts,W.H., 1943: A logical calculus of ideas immanent in nervous activity. Bull. Math. Biophys. 5, 115-133

McLeod,J.B., Fife,P.C., 1979: A phase plane discussion of convergence to traveling fronts for nonlinear diffusion. Preprint

Meinhardt,H., 1977: A model of pattern formation in insect embryogenesis. J. Cell. Sci. 23, 117-139

Melzak,Z.A., 1976: Mathematical ideas, modelling & applications. J. Wiley, N.Y. & L.

Mendall,L.M., Henneman,E., 1971: Terminals of single Ia fibers: location, denstiy and distribution within a pool of 300 homonymous motoneurons.J. Neurophys. 34, 171-187

Michel,A.N., Miller,R.K., 1977: Qualitative analysis of large scale dynamical systems. Academic Press, New York, London

Miller,R.K., 1971: Nonlinear Volterra integral equations. Benjamin Press, Menlo Park, California

Milsum,J.H., 1966: Higher order, nonlinear & spatially distributed models. In: Biological control systems analysis, pp. 209-239, McGraw Hill, New York

Mimura,M., Nishiura,Y., 1979: Spatial patterns for an interaction-diffusion equation in morphogenesis. J. Math. Biol. 7, 243-263

Morishita,I., Yajima,A., 1972: Analysis & simulation of networks of mutually inhibitory neurons. Kybernetik 11, 154-165

Motokowa,K., 1970: Physiologyand pattern vision. Springer-Verlag

Nagumo,J., Arimoto,S., Yoshizawa,S., 1962: An active pulse transmission line simulating nerve axon. Proc. IRE 50, 2061-2070

Nicolis,G., Prigogine,I., 1977: Self-organization in non-equilibrium systems. Wiley-Interscience, New York

Noble,D., 1966: Applications of Hodgkin-Huxley-equations to excitable tissues. Physiol. Rev. 46, 1-50

Nussbaum,R., 1973: Periodic solutions of some nonlinear autonomous functional differential equations II. J. Differential Equs.14, 368-394

Nussbaum,R., 1974: Periodic solutions of some nonlinear autonomous functional differential equations. Ann.Mat.Pura Appl.10, 263-306

Oguztöreli,M.N., 1975: On the activities in a continuous neural network. Biol. Cybern. 18, 41-48

Ortega,J.M., Rheinboldt,W.C., 1970: Iterative solution of nonlinear equations in several variables. Academic Press, New York and London

Oshima,T., 1969: Studies of pyramidal tract cells. In: Jasper,H., Ward,A., Pope,A. (eds.): Basic mechanisms of the epilepsis, pp. 253-261

Oster,G.F., Perelson, A.S., Katchalsky,A., 1973: Quart. Rev. Biophysics 6,1

Ostrowski,A.M., 1966: Solution of equations and systems of equations. Second edition. Acad.Press, New York

Othmer,H.G., 1977: Current problems in pattern formation. In:Lectures on Math. in the Life Sciences (S.A.Levin,ed.) 9, Amer. Math. Soc.,Providence

Paley,R.E.A.C., Wiener,N., 1934: Fourier transforms in the complex domain. Amer. Math. Soc. Colloquium Publications

Perkel,D.H., Mulloney,B., 1974: Motor pattern production in reciprocally inhibitory neurons exhibiting postinhibitory rebound. Science 185, 181-183

Pesin,J.B., 1974: On the behavior of a strongly nonlinear differential equation with retarded argument. Differentsialnye Uravnenija 10, 1024-1036

Poggio,T., Torre,V., 1978: A new approach to synaptic interactions. Lect. Notes in Biomath. 21,89-116

Rall,W., 1970a: Dendritic neuron theory and dendrodendritic synapses in a cortical system. In: Schmitt,P.O.(ed.): The neurosciences second study program. The Rockefeller Univ. Press, New York

Rall,W., 1970b: Cable properties of dendrites and effects of synaptic location. In: Anderson,P., Jensen,J.K.S.(eds.): Excitatory synaptic mechanisms. Universitets Forlaget, Oslo

Ratliff,F., Hartline,H.K., Miller,W.H., 1963: Spatial and temporal aspects of retinal inhibitory interaction. J. Opt. Soc. Am. 53, 110

Ratliff,F., Knight,B.W., Graham,N., 1969: On tuning and amplification by lateral inhibition. Proc. Nat. Acad. Sci. 62, 733-740

Ratliff,F., 1972: Contour & Contrast. Scientific American, 91-101

Ratliff,F. (ed.), 1974: Studies on excitation and inhibition in the retina. Chapman and Hall, London

Reichardt,W., 1961: Über das optische Auflösungsvermögen von Limulus. Kybernetik 1, 57-69

Reichardt,W., 1962: Theoretical aspects of neural inhibition in the lateral eye of Limulus. In: Information processing in the nervous system, Proc. Int. Union of Physiol. Sciences, Vol. III, 65-84

Reichardt,W., Mac Ginitie,G., 1962: Zur Theorie der lateralen Inhibition. Kybernetik 1, 155-165

Reiss,R.F., 1962: A theory and simulation of rhythmic behavior due to reciprocal inhibition in small nerve nets. Am. Fed. Inf. Process Soc. Proc. Spring Joint Computer Conference 21, 171-194

Rinzel,J., Keller,J.B., 1973: Traveling wave solutions of a nerve conduction equation. Biophys. J. 13, 1313-1337

Rinzel,J., Rall,W., 1974: Transient response in a dendritic neuron model for current injected at one branch. Biophys. J. 14, 759-790

Rinzel,J., 1975: Neutrally stable traveling wave solutions of nerve conduction equations. J. Math. Biol. 2, 205-217

Rinzel,J., 1975: Voltage transients in neuronal dendritic trees. Fed. Proc. 34, 1350-1356

Rinzel,J., 1976: Simple model equations for active nerve conduction and passive neuronal integration. In: Lectures on Mathematics in the Life Sciences, Volume 8, 89-124

Rinzel,J., 1978: Integration and propagation of neuroelectric signals. In: Studies in Math. Biology I. (Levin,S.A.,ed.), publ. by Math. Association of Amer.

Rothe,F., 1979: Some analytical results about a simple reaction-diffusion system for morphogensis. J. Math. Biol. 7, 375-384

Rothe,F., de Mottoni,P., 1979: A simple system of reaction-diffusion equations describing morphogenesis. I: Asymptotic behavior. Annali di Mat. Pura et Appl.,preprint

Sabah,N.H., Leibovic,K.N., 1972: The effect of membrane parameters on the properties of the nerve impulse. Biophys. J. 12, 1132-1144

Scott,A.C. 1977: Neurophysics. Wiley, New York

von Seelen,W., 1968: Informationsverarbeitung in homogenen Netzen von Neuronenmodellen. Kybernetik 5, 133 - 148

von Seelen,W., 1970: Zur Informationsverarbeitung im visuellen System der Wirbeltiere . Kybernetik 7, 43-60

von Seelen,W., Hoffmann,K.P. 1976: Analysis of neuronal networks in the visual system of the cat using statistical signals. Biol. Cyb. 22, 7-20

Shepard,G.M., 1974: The synaptic organization of the brain. Oxford Univ. Press, New York

Sobolewskii,P.E., 1965: Equations of parabolic type in a Banach space. Trans. A.M.S. 49, 1-62

Stanley,J.C., 1976: Simulation studies of a temporal sequence memory model. Biol. Cyb. 24, 121-137

Stein,R.B., 1967: The frequency of nerve action potentials generated by applied currents. Proc. R.Soc. Lond.B 167, 64-86

Stein,R.B., Leung,K.V., Mangeron,D., Oguztöreli,M.N., 1974: Improved neuronal models for studying neural networks. Kybernetik 15, 1-9

Stein,R.B., Leung,K.V., Oguztöreli,M.N., Williams,D.W., 1974: Properties of small neural networks. Kybernetik 14, 223-230

Stein,R.B., Oguztöreli, M.N. 1976: Tremor and other oscillations in neuromuscular systems. Biol. Cyb. 22, 147-157

Székely,G., 1965: Logical network for controlling limb movement in Urodela. Acta Physiol. Acad. Sci. Hung. 27, 285-289

Tokura,T., Morishita, I., 1977: Analysis and simulation of double-layer neural networks with mutually inhibiting interconnections.Biol.Cyb.25, 83-92

Tolhurst,D.J. Thompson, P.G., 1975: Orientation illusions and after effects: Inhibition between channels. Vision Research 15, 967-972

Tolkmitt,F.J., 1977: A computer simulation model of the afferent part of the visual foveation system. Biol. Cyb. 25, 195-203

Tuckwell,H.C., Miura, 1978: A mathematical model for spreading cortical depression. Biophysical J. 23, 257-276

Turing,A.M., 1952: The chemical basis of morphogenesis. Phil. Trans. Roy. Soc. (London) B 237, 37-72

Tyson,J.J., 1975: On the existence of oscillating solutions in negative feedback cellular control processes. J. Math. Biol. 1, 311-315

Tyson,J.J., Othmer,H.G., 1978: The dynamics of feedback control circuits in biochemical pathways. Progr. Theor. Biol. 6

Varga,R.S., 1962: Matrix iterative analysis. Prentice Hall, Inc. New Jersey

Varjú,D., 1962: Vergleich zweier Modelle für laterale Inhibition. Kybernetik 1, 200-208

Varjú,D., 1965: On the theory of lateral inhibition. In: E.R. Caianiello (ed.): Cybernetics of neural processes. Proceedings Inst. di Fisica theoretica, U. di Napoli

Varjú,D.,Pickering,S.G. 1970: Delayed responses of ganglion cells in the frog retina: The influence of stimulus parameters upon the discharge pattern. Kybernetik,6, 112-119

Varjú,D., 1977: Systhemtheorie. Springer Verlag, Berlin, Heidelberg, New York

Waterman,T.H., 1954: The functional relation between retinal cells and optic nerve in Limulus.J. Exp. Zool. 126, 252-257

Wevelsiep,K., 1977: Processing visual signals with nonlinear spatial filters, J. Math. Biol. 4, 81-99

Wiener,N., Rosenblueth,A., 1946: The mathematical formulation of conduction of impulses in a network of connected excitable elements, specifically in cardiac muscle. Arch. Inst. Cardiologia de Mexico 16, 205-256

Wilson,H.R., Cowan,J.D., 1973: A mathematical theory of the functional dynamics of cortical and thalamic nervous tissue. Kybernetik 13, 55-80

Wilson,H.R., Cowan,J.D., 1972: Excitatory and inhibitory interactions in localized populations of model neurons, Biophys. J. 12, 1-24

Winfree,A., 1978: Stably rotating patterns of reaction and diffusion. Theor. Chem. 4, 1-51

Zabreyko,P.O., Koshelev,A.I., Krasnoselski,M.A., Mihlin,S.F., Rakovshchik,L.S., Stet'senko, V. Ya., 1975: Integral equations, a reference text. Noordhoff Int. Publ. Leyden

LIST OF SYMBOLS

$N \subset M$	N is subset of the set M	
$x \in M$	x is element of the set M	
$\mathbb{R}$	the system of real numbers	
$\mathbb{R}_+$	the system of nonnegative real numbers	
$[a,b]$	the interval of all numbers x satisfying $a \leq x \leq b$	
(a,b)	the interval of all numbers x satisfying $a < x < b$	
$M \times N$	Cartesian product of two sets: $(x,y) \in M \times N \Longleftrightarrow x \in M, y \in N$	
$f: M \longrightarrow N$	symbol for a function f mapping the domain M into the range N	
$f \circ g$	composition of the functions f and g	p.9
w_t	derivative with respect to t	
w_{ss}	second derivative with respect to s	
$\{x: P\}$	set of all elements having property P	
δ	Dirac's delta function	p.2
$*$	convolution	p.4
h_{BA}, h_{ij}	temporal weight functions	p.2,6,11
$h(s_k, s_j)$	temporal weight function	p.13
g^*	Laplace transform of the function g	p.57
$D(\lambda)$	characteristic equation	p.57

S_{BA}	p.5,6	H_{ij}	p.5,27
T, T_B	p.9	$U = S \circ T$	p.9
T_{ij}	p.11	S_{ij}	p.11,27
$\bar{S}_i$	p.11	U_{ij}	p.12
$T(s_k, v)$	p.13	$S(s_k, s_j, x)$	p.13
$U(s_k, s_j, v)$	p.13	$m(x)$	p.16
e_j	p.27	$\bar{v}_i$	p.27

INDEX

Bio-mathematics

Managing Editors: K. Krickeberg, S. A. Levin

Springer-Verlag
Berlin
Heidelberg
New York

Volume 8

A. T. Winfree

The Geometry of Biological Time

1979. Approx. 290 figures. Approx. 580 pages
ISBN 3-540-09373-7

The widespread appearance of periodic patterns in nature reveals that many living organisms are communities of biological clocks. This landmark text investigates, and explains in mathematical terms, periodic processes in living systems and in their non-living analogues. Its lively presentation (including many drawings), timely perspective and unique bibliography will make it rewarding reading for students and researchers in many disciplines.

Volume 9

W. J. Ewens

Mathematical Population Genetics

1979. 4 figures, 17 tables. XII, 325 pages
ISBN 3-540-09577-2

This graduate level monograph considers the mathematical theory of population genetics, emphasizing aspects relevant to evolutionary studies. It contains a definitive and comprehensive discussion of relevant areas with references to the essential literature. The sound presentation and excellent exposition make this book a standard for population geneticists interested in the mathematical foundations of their subject as well as for mathematicians involved with genetic evolutionary processes.

Volume 10

A. Okubo

Diffusion and Ecological Problems: Mathematical Models

1980. 114 figures. XIII, 254 pages
ISBN 3-540-09620-5

This is the first comprehensive book on mathematical models of diffusion in an ecological context. Directed towards applied mathematicians, physicists and biologists, it gives a sound, biologically oriented treatment of the mathematics and physics of diffusion.

Lecture Notes in Biomathematics

Vol. 1: P. Waltman, Deterministic Threshold Models in the Theory of Epidemics. V, 101 pages. 1974.

Vol. 2: Mathematical Problems in Biology, Victoria Conference 1973. Edited by P. van den Driessche. VI, 280 pages. 1974.

Vol. 3: D. Ludwig, Stochastic Population Theories. VI, 108 pages. 1974.

Vol. 4: Physics and Mathematics of the Nervous System. Edited by M. Conrad, W. Güttinger, and M. Dal Cin. XI, 584 pages. 1974.

Vol. 5: Mathematical Analysis of Decision Problems in Ecology. Proceedings 1973. Edited by A. Charnes and W. R. Lynn. VIII, 421 pages. 1975.

Vol. 6: H. T. Banks, Modeling and Control in the Biomedical Sciences. V, 114 pages. 1975.

Vol. 7: M. C. Mackey, Ion Transport through Biological Membranes, An Integrated Theoretical Approach. IX, 240 pages. 1975.

Vol. 8: C. DeLisi, Antigen Antibody Interactions. IV, 142 pages. 1976.

Vol. 9: N. Dubin, A Stochastic Model for Immunological Feedback in Carcinogenesis: Analysis and Approximations. XIII, 163 pages. 1976.

Vol. 10: J. J. Tyson, The Belousov-Zhabotinskii Reaktion. IX, 128 pages. 1976.

Vol. 11: Mathematical Models in Medicine. Workshop 1976. Edited by J. Berger, W. Bühler, R. Repges, and P. Tautu. XII, 281 pages. 1976.

Vol. 12: A. V. Holden, Models of the Stochastic Activity of Neurones. VII, 368 pages. 1976.

Vol. 13: Mathematical Models in Biological Discovery. Edited by D. L. Solomon and C. Walter. VI, 240 pages. 1977.

Vol. 14: L. M. Ricciardi, Diffusion Processes and Related Topics in Biology. VI, 200 pages. 1977.

Vol. 15: Th. Nagylaki, Selection in One- and Two-Locus Systems. VIII, 208 pages. 1977.

Vol. 16: G. Sampath, S. K. Srinivasan, Stochastic Models for Spike Trains of Single Neurons. VIII, 188 pages. 1977.

Vol. 17: T. Maruyama, Stochastic Problems in Population Genetics. VIII, 245 pages. 1977.

Vol. 18: Mathematics and the Life Sciences. Proceedings 1975. Edited by D. E. Matthews. VII, 385 pages. 1977.

Vol. 19: Measuring Selection in Natural Populations. Edited by F. B. Christiansen and T. M. Fenchel. XXXI, 564 pages. 1977.

Vol. 20: J. M. Cushing, Integrodifferential Equations and Delay Models in Population Dynamics. VI, 196 pages. 1977.

Vol. 21: Theoretical Approaches to Complex Systems. Proceedings 1977. Edited by R. Heim and G. Palm. VI, 244 pages. 1978.

Vol. 22: F. M. Scudo and J. R. Ziegler, The Golden Age of Theoretical Ecology: 1923–1940. XII, 490 pages. 1978.

Vol. 23: Geometrical Probability and Biological Structures: Buffon's 200th Anniversary. Proceedings 1977. Edited by R. E. Miles and J. Serra. XII, 338 pages. 1978.

Vol. 24: F. L. Bookstein, The Measurement of Biological Shape and Shape Change. VIII, 191 pages. 1978.

Vol. 25: P. Yodzis, Competition for Space and the Structure of Ecological Communities. VI, 191 pages. 1978.

Lecture Notes in Biomathematics

Vol. 26: M. B. Katz, Questions of Uniqueness and Resolution in Reconstruction from Projections. IX, 175 pages. 1978.

Vol. 27: N. MacDonald, Time Lags in Biological Models. VII, 112 pages. 1978.

Vol. 28: P. C. Fife, Mathematical Aspects of Reacting and Diffusing Systems. IV, 185 pages. 1979.

Vol. 29: Kinetic Logic – A Boolean Approach to the Analysis of Complex Regulatory Systems. Proceedings, 1977. Edited by R. Thomas. XIII, 507 pages. 1979.

Vol. 30: M. Eisen, Mathematical Models in Cell Biology and Cancer Chemotherapy. IX, 431 pages. 1979.

Vol. 31: E. Akin, The Geometry of Population Genetics. IV, 205 pages. 1979.

Vol. 32: Systems Theory in Immunology. Proceedings, 1978. Edited by G. Bruni et al. XI, 273 pages. 1979.

Vol. 33: Mathematical Modelling in Biology and Ecology. Proceedings, 1979. Edited by W. M. Getz. VIII, 355 pages. 1980.

Vol. 34: R. Collins, T. J. van der Werff, Mathematical Models of the Dynamics of the Human Eye. VII, 99 pages. 1980.

Vol. 35: U. an der Heiden, Analysis of Neural Networks. X, 159 pages. 1980.